ON PETROCULTURES

Energy and Society
Brian Black, Series Editor

On Petrocultures

Globalization, Culture, and Energy

Imre Szeman

West Virginia University Press
Morgantown 2019

ISBN
Cloth 978-1-946684-87-5
Paper 978-1-946684-88-2
Ebook 978-1-946684-89-9

Library of Congress Cataloging-in-Publication Data
Names: Szeman, Imre, 1968- author.
Title: On petrocultures : globalization, culture, and energy / Imre Szeman.
Description: First edition. | Morgantown : West Virginia University Press, [2019] |
        Series: Energy and society | Includes bibliographical references and index.
Identifiers: LCCN 2018040462| ISBN 9781946684875 (cloth) | ISBN 9781946684882
        (pbk.)
Subjects: LCSH: Petroleum industry and trade--Social aspects. | Petroleum industry
        and trade--Political aspects. | Globalization. | Neoliberalism. |Petro-
        leum--Social aspects. | Power resources--Social aspects.
Classification: LCC HD9560.5 .S994 2019 | DDC 338.2/728--dc23
LC record available at https://lccn.loc.gov/2018040462

Cover design by Than Saffel / WVU Press

for Pulpo, with love

CONTENTS

Acknowledgments

The essays collected here represent research and writing that I carried out over close to two decades. I could not have done any of this work without the invaluable support of friends and colleagues, whose influence and impact can be found across these pages. There are many people to thank and honor—so many that I know I will have missed people whose names deserve to be listed here. For those who might have expected to find themselves on these pages but are missing, my apologies in advance!

At Duke University, I'd like to thank Michael Hardt, Fredric Jameson, and Valentin Mudimbe; I'd like, too, to honor the memory of the late Arif Dirlik and Ted Davidson, members of my doctoral thesis committee who weren't afraid to tell me what they really thought.

At McMaster University, thank you to my amazing colleagues in the Department of English and Cultural Studies and the Institute on Globalization and the Human Condition: Daniel Coleman, William Coleman, Henry Giroux, Don Goellnict, Jasmin Habib (now my colleague again at Waterloo), Max Haiven, Tim Kaposy, Grace Kehler, Matt MacLellan, Susie O'Brien, Mary O'Connor, and Lorraine York. I miss Mac even to this day.

At University of Alberta, thanks to my colleagues in the Department of English and Film Studies, including Karyn Ball, Mo Engel, Nat Hurley, Gail Faurschau, Eddy Kent, Sourayan Mookerjea, Michael O'Driscoll, Julie Rak, Mark Simpson, Stephen Slemon, Terri Tomsky, and Heather Zwicker. The graduate students and postdoctoral fellows with whom I worked at Alberta are poised to change the world with their ideas. They include Lynn Badia, Brent Bellamy, Adam Carlson, Jeff Diamanti, Derek Gladwin, Dan Harvey, David Janzen, Jordan Kinder, Sean O'Brien, Kurt Pabst, Valérie Savard, and Lucie Stepanik. Thanks for giving me the opportunity to work with all of you.

At the University of Waterloo, I want to thank in particular Angela Carter, Beth Coleman, Rob Danisch, and Jennifer Simpson for making me feel at home in my new institutional home. Jessica Hawkes at

Waterloo proved to be an invaluable research assistant in the summer of 2018.

At the Banff Centre for Arts and Creativity, a thank you to Kitty Scott (now at the Art Gallery of Ontario) and to Brandy Dahrouge, who is a large part of why Banff Research in Culture works year in and year out.

To Nicholas Brown, for being a wonderful colleague, collaborator, and friend—thanks.

Thanks to Derek Krisoff at West Virginia University Press, who believed in this book project from the very beginning. I look forward to working with you on other projects in the future.

Thank you to all those who are important to me, scattered near and far, with whom I wish I could spend more time: Marija Cetinic (Amsterdam); Zoran Vuckovac and Danijela Majstorović (Banja Luka); Fengzhen Zheng (Beijing); Eva Boesenberg (Berlin); Jessie Labov and György Tury (Budapest); Shaobo Xie (Calgary); Rachel Havrelock (Chicago); Graeme Macdonald (Glasgow); Dominic Boyer and Cymene Howe (Houston); Amy De'Ath (London); Patrick McCurdy and Joseph Szeman (Ottawa); Darin Barney, Marty and Kinny Kreiswirth, Sebastian Thomson-Jagoe, and Will Straw (Montreal); Danine Farquharson (St. John's); Adrienne Batke, Art Blake, Elspeth Brown, David Carter, Andrew Herman, Andrew Johnson, Smaro Kamboureli, Lauren Kirschner, Caroline Langill, Justin Manuel (owner of Black Cat Espresso Bar), Tanner Mirrlees, and Sarah Sharma (Toronto); Todd Dufresne (Thunder Bay); Markus Heide (Uppsala); Ruth Beer, Richard Cavell, Shane Gunster, Penelope Hutchison, Liam Jagoe, Michael Ma, and Stuart Poyntz (Vancouver). Let's scheme up some new projects together!

A million thanks to Sarah Blacker, Andrew Pendakis, and Justin Sully, who have been so much more than friends and so much more than colleagues to me. Their energies and enthusiasms have made my own experiences and ideas—heck, my whole life—richer and deeper.

Thanks to Caleb Wellum, for his exceptional care in putting the finishing touches on the manuscript. Caleb: this book was made far better due to your input and advice.

Thanks, too, to Paul Schellinger for his fine work in readying the book for publication.

And finally, to my best friend, my intellectual companion, my sharpest critic and strongest supporter, and the love of my life, Eva-Lynn Jagoe. If I can't wait to see what the future brings, it's because I know that I'll get to see it with her by my side.

Introduction

On Petrocultures

I really started to notice it in fall 2017. Some of the places that I visited on a daily basis—a park to which I take my puppy for a scamper, a coffeehouse that I visit to do some writing—began to buzz with ideas I was not used to hearing in downtown Toronto. On a late-November day when the temperature reached +17 C (63 F), Jack-the-terrier's owner remarked: "Did you notice that they've started calling it 'climate change' instead of 'global warming'?" He suggested that maybe we should just enjoy warm winter temperatures instead of worrying about them. Jack's owner, an engineer who works in the Arctic from time to time, began to tell me about what he had encountered on trips to the North. I thought it was going to be about melting ice and disappearing glaciers. He spoke instead about core samples and what their findings show. Maybe they *didn't* indicate that the planet was warming after all. He held up his hands in protest, "Don't get me wrong, I know that there's climate change. I'm just saying . . ."

This turned out to be just the first such shift of attitudes about climate that I encountered. The Black Cat on College Street attracts a mix of patrons, from graduate students struggling with their theses to the owners of stores located nearby who pop in for an espresso. One person who visits multiple times a day is an older, small-time lawyer who holes up in a corner with friends. They argue affably with one another about every topic under the sun. The corner table at The Black Cat is sometimes a site of discussion about the meaning of life, whether it's possible to write about something you haven't experienced (one of the crew writes fantasy novels), changes in the way that the sexes interact, or the fate of local sports teams. Their comments are rarely objectionable, even if the facts and figures they sometimes espouse are questionable—a Google Search view of the world. For the most part, these men are aware that people are

1

shaped by their environments, and their headshaking about the world tends to be about the injustices inflicted by the rich and powerful, or about incompetencies characterizing local politics.

As of late, however, my eavesdropping at The Black Cat has left my blood boiling. While these older, liberal men talk about a range of topics, these days they inevitably end up on the subject of the environment. What they talk about isn't how bad things are, or how little action there appears to be on global warming in Canada.[1] Rather, they feel misled about climate change. They're pretty sure that things are not as bad as experts say. They sense a conspiracy afoot. One of the men returns again and again to two points of evidence to back up these claims. The first concerns results from the oldest continuously operating temperature gauge in the world. He claims that this gauge shows that, contrary to everything that scientists have been telling us, the temperature of the planet is *decreasing*.[2] The second concerns claims made in Al Gore's documentary, *An Inconvenient Truth* (2006). A little over a decade later, the man in the corner insists excitedly that many of Gore's claims about climate change have been irrefutably proven to be "completely false!"

I've talked to Jack's owner, cautiously pointing out, for instance, that if there has been a shift from the language of global warming to climate change, then it's a trick and a con—a way of shaving off the rough edges of the need for significant and immediate action to address the warming of the planet and producing instead a situation that seems accessible to policy prescriptions and technological solutions, while also introducing a hint of doubt (in just what way is climate changing?). I've drawn attention to the repeated findings and pronouncements made by armies of scientists attesting to the reality of global warming—including recent reports about the Arctic.[3] Though he sees himself as a concerned Canadian and good citizen, Jack's owner remains unconvinced. He thinks that there is something fishy going on, though he's at least willing to concede that he might be wrong about just what that is.

As for The Black Cat patrons? I imagine interrupting their conversation. I want to correct their assumptions, challenge their evidence, and convince them that they need to take all manner of environmental crises more seriously. Luckily, I think, I've hesitated up until now. I have a sense the conversation would go awry rather quickly, drifting into *ad*

hominem attacks about the fact that I drive a car or take flights to conferences and events, or even about the choices I make about coffee—you
name it (in my defense, Canada does not produce a lot of locally grown
coffee). I want to tell them that scientists have, almost without exception,
made it clear that we are facing a planetary emergency.[4] But they would
take small and slight exceptions as clear challenges to the rule I was
trying to argue. In an era of fake news, it has become easy to constitute
challenges to expert opinion and authority;[5] in an age of conspiracy, the
narrative of the scientific voice in the wilderness—the lone wolves who
aren't convinced about global warming—is often going to be attractive
to people who don't know, or don't want to know, what's going on.[6]

As the growth of climate skepticism shows, one of the greatest social
and political challenges of the present moment is how to effectively
communicate about global warming.[7] This is a challenge less of available
data than of rhetoric and representation, as many critics have pointed
out.[8] Texts in different genres have hoped to have the intended outcome of Gore's film: to alert readers or viewers about a problem so
that they might effect a change in practice. These documents include
everything from the chain of reports issued by the United Nations'
Intergovernmental Panel on Climate Change to information one can
find on the website of the Union of Concerned Scientists, and from
Pope Francis's *Laudato Si'* (*Encyclical on Climate Change and Inequality,*
2015) to the actual text of the Paris Climate Agreement (2015), as well
as a range of texts—in film and fiction—that have struggled with the
representation of a world impacted by global warming.[9] The essays
brought together in this book are animated by a hope and desire to
create new ways of talking about the challenges of global warming—
and about other long accepted verities as well, including nationalism,
globalization (another favorite topic of the coffee shop corner table),
and the operations of twenty-first-century capitalism, which have been
propelled by new categories of acting and being such as creativity and
entrepreneurship.

One of the topics at the heart of this book is how we relate—or
fail to relate—to energy. When it comes to talking with the people
I encounter on a daily basis in the communities in which I live, I'm
tempted to challenge their views by playing the role of professor. Surely

confronting them with the evidence about climate change accumulated by technical and scientific expertise will get them to see things differently! But contemporary subjects live exceptionally complex lives that aren't necessarily governed by rational choices formulated around the careful consideration of scientific evidence. The inevitable pressures of work and life, combined with the fragmented manner in which all of us cobble together information on any almost any subject whatsoever, means that as often as not we are governed by those hegemonic, conservative truisms to which Roland Barthes draws our attention in *Mythologies*.[10] Pointing didactically to hyperobjects like the climate or global warning isn't likely to generate a change in my fellow citizens.[11] But I've found that engaging them with a story about energy, about our demand for it and the limits of its availability, and about the consequences of its use—at once geopolitical and environmental—shifts the frame of reference; it moves the analysis from the abstraction of the climate to the reality of the gas tank. It enables us to have a discussion, too, about one of the things too often hidden in regular discussions of global warming: the past, present, and future of oil capitalism and the petrocultures it has generated.

* * *

On Petrocultures brings together twelve essays I have written over the past two decades. They cover a diverse range of topics—from a defense of the concept of "national allegory" in relation to postcolonial fiction to the state and fate of cultural theory in the context of globalization, and from neoliberal refigurations of autonomy and culture (via the celebration of entrepreneurship and creative economies) to essays that played a role in shaping the field of "energy humanities."

It would be a stretch to say that these essays are all part of a single, coherent project—that is, that my argument about the importance of national allegory leads directly to the elaboration of "pipeline theory" (in one of the previously unpublished essays in this collection). Even so, as distinct as their topics might be, they *are* animated by a shared set of concerns. These essays are about life lived in late capitalism, the operations and circulations of key concepts we use to comprehend and narrate our collective histories, and the ways in which our intellectual and political commitments to these concepts aid or impede social and political justice. What did the concept of the nation do to the radical political energies of postcolonial cultures and societies? How did the narrative of post-Soviet globalization cover up, cover over, and legitimate intensified practices of neoliberalism and resource extraction? What are the cultural and social implications of discourses of creativity (as foregrounded in Richard Florida's concept of "creative cities") and entrepreneurship, especially with respect to critiques of capitalism and arguments for the creation of new collectivities? In each case, what I have been concerned with is an understanding of the broad political implications of the vocabulary through which contemporary capitalism makes sense of itself, and through which we have been invited to make sense of it, too.

As throughout my work (beginning with *Zones of Instability*), there is here a fascination with the delimitation of space by nations, cultures, and borders.[12] The nation has to be counted as one of the most extraordinarily powerful fictions of the modern era—a way of connecting sovereignty to geography that has come to be taken as the accepted, principal mode of political power. Its mechanisms of geo-sovereignty have, since at least Johann Gottfried von Herder, also acted as containers for the practices, activities, and beliefs known as "culture," which in turn have helped to legitimate national sovereignty by noting and naming the distinctions between the people living in different nations.[13] At least from one perspective, the advent of post-Soviet globalization promised to bring about an end to national-cultural struggles, via the birth of a cosmopolitan world positioned as the next step in a never-ending process of Kantian maturation. Now we all belonged to one world, one global culture, even if we still lived within our separate nations. From the beginning, however, globalization was as much of a fiction as nationalism, flattening the world out only for trade, commerce, and the movement of money, and dividing the world into (in Zygmunt Bauman's memorable taxonomy) "tourists" and "vagabonds," those who could travel freely across borders and those who couldn't (or could only do so illegally).[14] Globalization was imagined as the end of nations or national-cultures only for elites who stood to benefit from this redefinition of borders. The practices and policies of neoliberalism that gave life to globalization have created new discourses of subjectivity—of being and belonging, acting and reacting, in which personal autonomy and cultural identity are linked in increasingly complicated ways with the yawning gap in the social left by the retreat of the state into the market. One core group of essays in this book maps the complex, often violent games of power, culture, and borders that continue to be played out across the globe, to the benefit of some and the detriment of the vast majority of the planet's inhabitants.

Another group of essays collected here examines the epistemology and social ontology of energy. When I began actively thinking about fossil fuels, what struck me with enormous force was the absence of a concerted exploration of the importance of energy for modernity. Fossil fuels were hiding in plain sight—an obvious force in wars, geopolitics, infrastructure developments, and almost everything associated with modern "progress" (from the expansion of technology to the growth of democracy). And yet fossil fuels and energy were a missing element of critical scholarship. They were missing, too, from the literature and culture of modernity, from those novels and films that we rely on to capture the unconscious forces and stresses shaping the social. When I would make these claims at conferences or in talks, or in my written work, most people would grasp immediately the enormous absence to which I was pointing. Some, however, would claim that I was overstating things, and would point to the importance of fuel in shaping novels (for instance), everything from Herman Melville's *Moby Dick* (whale oil, 1851) to Upton Sinclair's *Oil!* (the stuff of the book's title, 1926–27) to Karel Čapek's remarkable *The Absolute at Large* (free energy, 1922). This really is a case, however, where the few exceptions prove the larger rule: there has been a marked failure to theorize energy, its significance, and its consequences in any concerted manner.

It is hard not to narrate the geopolitical history of the United States in relation to fossil fuels. Access to oil is at the center of its rise to the role of superpower in the twentieth century, and the drive for ever more of the stuff is key to its ongoing manipulation of Middle Eastern politics. The current Trump administration is nothing if not a petro-cultural government, whether measured by the number of former oil executives occupying senior positions in the administration, or by its fetish for coal and fantasies of a mid-century white America that was made possible (in part) by enormous amounts of dirty energy. Amitav Ghosh once asked why there was so little attention in US fiction to the country's geopolitical misadventures in the Middle East.[15] The same question can be asked of other fictions and cultural forms, to probe why the substance at the heart of so many geopolitical misadventures has been missing everywhere and in our accounts of everything—a gap with enormous consequences with which we are only now beginning to grapple. We moderns are creatures of fossil fuels (if to different degrees in different places in the world), a fact we have largely chosen to forget in our accounts of our histories, politics, and culture.

The area of research now known as the "energy humanities" has, over the past decade or so, begun to fill in some of the missing pieces of the puzzle of energy in contemporary culture.[16] My own work has tried, in part, to make sense of whether there is something about fossil fuels that has made them difficult to represent, which might explain why an element of such importance to the shape and form of modernity has been missing from characterizations of the modern. I've approached this question in two distinct, if related, ways. The first has been to assess the socio-political and cultural reasons for the under-representation of energy in general, and fossil fuels more specifically, in literature, film, and other forms of cultural representation. This mode of analysis draws attention to the character of oil's presence or absence in the political imaginaries of modernity, and tries to offer reasons for why energy hides in plain sight. My second approach has been to ask deeper questions about the ontology of oil and energy, and to consider what this ontology implies for how we have characterized oil capitalism and constituted challenges to it. In both cases, I've wanted to understand if there is something about the nature of fossil fuels (oil *qua* oil) that makes them resistant to fully naming and explaining their essence and their essential role in shaping modernity. Do representations of fossil fuels—those narratives and images that do in fact exist—miss the mark when they fail to capture how and why this is a substance that *cannot* be represented? In their eagerness to deal with a missing element, do they show it off too quickly, too didactically, and so fail to actually address the reasons for its absence in the first place? How, then, might one make sense of such a substance, and bring to light the political and environmental consequences of our having shaped every aspect of civilization around it?

The process of coming to grips with the unique character of fossil fuels, epistemologically as well as ontologically, has been a difficult one, with at times an unclear trajectory. While there is a large and rich body of research that explores fossil fuels and society, especially in the fields of anthropology, history, and (more recently) literary criticism, when I began my reflections I found only a few essays and books that spoke directly to my particular concerns. This small cluster of work is book-ended by Ghosh's insightful critical writing—beginning with "Petrofiction: The Oil Encounter and the Novel" (1992) and ending with *The Great Derangement: Climate Change and the Unthinkable* (2016)—which asks big-picture questions about the incapacities of fiction to name ourselves in relation to our petrocultures, and challenges us to develop modes of representation that might yet help us more fully grapple with the substance animating modernity.[17]

A handful of other texts have been critically important as well. I still know of no book that better articulates the implications of the power unleashed by fossil fuels for human ontologies, epistemologies, and ethics than Allan Stoekl's *Bataille's Peak: Energy, Religion, and Postsustainability* (2007); in my opinion, this is a work that has yet to receive the full critical response it deserves, an engagement that would generate new explorations of the deep connection between subjectivity, community, and energy (I plan to devote critical time and energy to this book in the near future). Dipesh Chakrabarty's "The Climate of History: Four Theses" (2009) and Timothy Mitchell's *Carbon Democracy: Political Power in the Age of Oil* (2011) offer profound re-narrations of modernity by highlighting the element all too often missing from our histories: energy. As these thinkers have done with history in relation to energy, Jeff Diamanti has provocatively done with (among other things) the built environment of modernity, which he describes as having been shaped by a process of "energy deepening," an ever-expanding, recursive commitment to fossil fuels that might well have made it extraordinarily difficult to generate alternatives to petroculture. Stephanie LeMenager and Jennifer Wenzel have each helped me to see how foregrounding fossil fuels can generate profound new ways of understanding cultural narratives. LeMengaer's essay, "Petro-Melancholia: The BP Blowout and the Arts of Grief," alerts us to the affective dimensions of our relation to fossil fuels, including the genres, practices, and mechanisms we use to critique this relation; our critical texts, too, are marked by their emergence in a period in which energy was imagined as infinitely available, which means that critique is also filled with and defined by the energies of petromodernity. Finally, in addition to her many other contributions to the field, Wenzel's introduction to *Fueling Culture* remains the single best overview of the questions and concerns animating energy humanities; anyone interested in this field would find this a great place to start.[18]

I describe my work on oil—both the essays included in this book and my broader work on energy—as an attempt to develop a *critical theory of energy*. As surely as all of the other systems and practices interrogated by a critical theory of society, the processes associated with and generated by our use of energy produce social possibilities and limits that need to be thoroughly explored. From its origins, the aim of critical theory was to imbue social analysis with an explicitly political intent. Critical theory would profess not only to understand the forces and processes that shaped social forms and practices, but to do so for emancipatory purposes. It's hard not to hear the echo of Marx's eleventh thesis from *Theses on Feuerbach* in Max Horkheimer's description of the activity of Frankfurt School critical theory: "Its goal is man's emancipation from slavery."[19] In addition to its explicit political goal, the theory and analysis undertaken in critical theory would also be alert to the limits of theory as such, by being permanently attentive and attuned to the positivisms and utilitarianisms to which social analyses are vulnerable. Critical theory determinately jettisons the political fiction that social science might be able to mirror the disinterest and objectivity of the natural sciences. When we speak today of critical theory in a more generic sense—that is, as not just the direct outgrowth of the work of the Frankfurt School, but as a practice that more commonly goes by the name "theory"—it is this that I think we have in mind: an alertness and awareness of the damage that concepts can do when heuristics are mistaken for (whether accidentally or deliberately) facts of nature.

Given this, exactly what might it mean to develop a critical theory of energy? Energy should not be taken as just a *topic* of study for the human sciences—another field to add (say) to the environmental humanities, or a needed reminder of the resource politics that constitutes an essential component of colonialism and postcolonialism. The new attention to energy that has been promulgated by the energy humanities is intended to unnerve the continuing legibility of the study of history, politics, philosophy, and literary and cultural studies, as presently practiced. The critical work involved in the energy humanities goes well beyond the "ta da!" of revealing a missing component of the various core elements of the narrative of the modern—of democracy, belonging, and community, of colonialism and postcolonialism, and indeed, even of the constitution of subjectivity. If this was all that looking at energy accomplished, there would be little left to do once the initial reveal took place. In truth, what the emergence of energy in the field of the human sciences demands is not just a slight amelioration of critical vocabularies, a nip-and-tuck addition of energy to the discourses we already have, but a wholesale refashioning of these vocabularies and their presumed objects of study.

The modern subject, for example, has had her capacities radically redefined by cheap energy. The petrocultural subject's life is configured around the energies of millions of years of dead matter; what she understands as banal quotidian reality is in fact a bending and stretching of time to give the subject powers of movement, vision, and knowledge akin to that of the demigods found in ancient myth. To the powers of the unconscious mapped by Freud and the powers of political economy interrogated by Marx, we need to add the capacities of energy, which inhabit (and shape and form) both these spaces and still others we've yet to fully probe. In short, a critical theory of energy insists that *adding energy to the mix of our analyses on any subject whatsoever forces a refashioning of the theories of the forces that animate the social and the subjects within it.*[20]

The fundamental challenge that energy poses to theory is one that scholars are slowly beginning to recognize. What is perhaps less evident are the political and emancipatory forces that emerge from critical attention to energy. Remember Horkheimer's words: the goal of critical theory is emancipation from slavery. One way in which some writers on energy have tried to insist on its social import and significance is by linking it *directly* to the practice of slavery. As a way of changing attitudes toward our current levels of energy use, "emotionally as well as intellectually," Jean-François Mouhot points out, "if we all wanted to benefit from our current lifestyles without any fossil fuels, we would need to employ several dozen people working full time for us."[21] The energy of fossil fuels allows us to do things we could not otherwise do; only by drawing on the energies of other bodies in servitude to us could we approximate the now taken for granted powers of the modern subject. The link made to slavery suggests that the critical aim of the energy humanities is to get us to abandon fossil fuels in favor of other, greener forms of energy, and so interrupt the operations of modernity as a formation in which we live way beyond the limits of our physical bodies and collective means. Mouhot has still other reasons for drawing the connection between energy and slaves. He worries that there is a real risk of a return to slavery in the future if dependence on fossil fuels is not addressed in the present, as elites try to maintain fossil-fuel comforts and lifestyles even in a post–fossil fuel world. However powerful such analyses of our fossil-fuel servitude might be, using oil is *not* the same as using slaves; the appeal by Mouhot and others to the notion of "energy slaves" is, at best, an attempt to gesture toward an ethics via a somewhat offensive allegory. One of the side effects of this view of fossil fuels is that it has inadvertently helped shape a strongly held belief that a shift to wind and solar power occasions a more general expansion of social justice—a completely unsubstantiated view of how energy and social possibility are linked. The real political and emancipatory force of the energy humanities lies elsewhere.

Our use of fossil fuels is the single biggest source of human-produced CO_2. In other words, it is in large part our commitment to our petrocultures, both conscious and unconscious, that has re-shaped the global environment, and in a very short period of time. Fossil-fuel use and CO_2 production has expanded precipitously since World War II, a period referred to by some critics as the Great Acceleration.[22] During this period, expanded uses of fossil fuels enabled increases in population (through, for example, industrial agriculture and the widespread use of fertilizers), which in turn increased the demand for fossil fuels, and so on, generating a human footprint of astonishing proportions on the planet.[23] In order to address the impact of fossil fuels on the environment, it is clear that we will need to undergo an energy transition this century—a shift from dirty energy to clean, renewable, and sustainable forms of energy. The consensus among scientists of the need for energy transition is matched today by the majority of the world's policy makers: documents like the Paris Agreement are not only about limiting CO_2 output, but also about enabling energy transition. To date, very few countries have actually engaged in the processes needed to undergo energy transition; the most well known and most successful of these is Germany's impressive *Energiewende* (energy transition), which in short order has managed to make renewables the dominant energy source in the country.[24]

Almost without exception, discussions of energy transition make a presumption that needs to be challenged. Fossil fuels have played an essential role in activating and enabling the operations of capitalism, and all that comes with it: profit over people, ferocious extractivist practices, and social and political injustice. For the most part, both existing and planned practices of energy transition presume the persistence of contemporary capitalism and the forms of neoliberal governance that have accompanied it.[25] Current models of energy transition *do not* imagine that, in order for it to be successful, there must also be transitions in culture, society, and politics, too. This should come as little surprise. It is not only the self-interest of the status quo that is at work in imagining that a capitalism developed in conjunction with the easy-to-access (until recently) and cheap energy of fossil fuels could continue long into the future, with nary a hiccup or misstep, but now energized by solar panels and wind farms (with an added benefit of a desperately sought after injection of value through the introduction of new forms of energy).[26] The gaps and absences that the energy humanities have noted in conceptions of our social and cultural past remain alive, as well, in most articulations of our political futures. From the perspective of the status quo, energy doesn't seem to be anything other than the input we need to make things run. So why couldn't there thus be a renewable energy capitalism, one that would continue to make profit without harming the planet?[27]

I believe the period of energy transition offers a political opening—an unexpected one. The energy of fossil fuels has given rise to a specific way of life. Oil has shaped our communities in ways that have produced an abiding loyalty to the substance; it has generated a violent geopolitics organized around access to and control over the "black gold" that fuels the planet. This way of life is one that is neither especially well-loved nor actively defended by the planet's human inhabitants. The majority of us have no real commitment to the inequalities and injustices that oil capitalism and its political mechanisms have promulgated, and certainly don't wish to destroy the environment. As fossil-fuel capitalism begins to shift to other forms of energy—reluctantly and haphazardly, as evidenced by the last stand of fossil-fuel capitalism in the Trump administration—it seems to me that the possibility of other transitions opens up. One of the key political struggles of this century will be whether an energy transition can be achieved without other forms of transition, or whether the profound dislocation of moving away from fossil fuels, in the context of a growing awareness of global warming, will generate social and cultural transitions, and political ones, too. Dipesh Chakrabarty writes: "The mansion of modern freedoms stands on an ever-expanding base of fossil-fuel use. Most of our freedoms so far have been energy-intensive."[28] The emancipatory struggle announced by a critical theory of energy is to develop a world not only beyond the dictates of oil capitalism, but also one whose freedoms aren't the outcome of the use of massive amounts of energy. The carbon democracy whose development Timothy Mitchell outlines is a democracy in name only; and many of our freedoms to date have been little more than a fantastical deification of the human, giving some of us powers and capacities, but operating on the borrowed time of fossilized bodies, at the expense of the world we inhabit.

* * *

This call to arms for new freedoms and new modes of democracy is hardly new; indeed, it defines the genre of left politics in a moment of political impasse and interregnum. What I hope *is* new is that a concern with energy links the possibility and nature of freedom and democracy directly and deeply to the physical and the material. It places the language of rights in conjunction with the vocabulary of kilojoules and British thermal units.[29] This link of freedom with physical power raises an issue that has rarely been analyzed, but which constitutes yet another dimension of critical theory of energy: energy justice.[30]

The issues and challenges associated with energy justice arise in, of all places, Raymond Williams's "Culture Is Ordinary" (1958). This essay is most frequently cited as a powerful rejoinder to attempts to denigrate the culture of ordinary people. Williams's critique is directed as much to the anxieties about mass culture expressed by Max Horkheimer and Theodor Adorno in *Dialectic of Enlightenment* (1944) as it is to the affirmation of the Arnoldian view that ordinary popular literature tends to "teach down to the level of inferior masses."[31] To claim that culture is ordinary is to say that culture is a space of life and experience, a resource for groups and individuals, which is diminished neither by the existence of mass, commercial culture, nor by the use of culture to legitimate social divisions via games of distinction and the exercise of class power (as it is in the use of culture in relation to the creative culture of entrepreneurial society). "I believe the central problem of our society, in the coming half-century," Williams writes, "is the use of our new resources to make a good common culture; the means to a good, abundant economy we already understand."[32]

More than half-a-century later, it comes as a surprise to hear Williams suggest that we already understand the means to "a good, abundant economy." The continued expansion of global per capita GDP since 1958 certainly speaks to an "abundant" economy, even if this plenty has largely been retained by elites at the tiptop of the economic ladder. The good economy here is imagined in relation to the Keynesian compromises of the postwar era, which create a situation starkly different from the horrors outlined just two decades earlier in George Orwell's *The Road to Wigan Pier* (1937). It is also connected to the capacities introduced into daily life by industry and technology:

At home we were glad of the Industrial Revolution, and of its consequent social and political changes. True, we lived in a very beautiful farming valley, and the valleys beyond the limestone we could all see were ugly. But there was one gift that was overriding, one gift that at any price we would take, the gift of power that is everything to men who have worked with their hands. It was slow in coming to us, in all its effects, but steam power, the petrol engine, electricity, these and their host of products in commodities and services, we took as quickly as we could get them, and were glad. I have seen all these things being used, and I have seen the things they replaced. I will not listen with patience to any acid listing of them—you know the sneer you can get into plumbing, baby Austins, aspirin, contraceptives, canned food. But I say to these Pharisees: dirty water, an earth bucket, a four-mile walk each way to work, headaches, broken women, hunger and monotony of diet. The working people, in town and country alike, will not listen (and I support them) to any account of our society which supposes that these things are not progress: not just mechanical, external progress either, but a real service of life. Moreover, in the new conditions, there was more real freedom to dispose of our lives, more real personal grasp where it mattered, more real say. Any account of our culture which explicitly or implicitly denies the value of an industrial society is really irrelevant: not in a million years would you make us give up this power.[33]

In "Culture Is Ordinary," the good economy is made possible by the expansion of industrial society—by its mechanisms and technologies, which enrich ordinary life by opening up freedoms and opportunities never before available to working people. The prospects that one finds in ordinary culture as a result of industrial society is linked, too, to "the gift of power"—"steam power, the petrol engine, electricity, these and their host of products in commodities and services." The introduction of power led to the introduction of modern freedoms: "more real personal grasp where it mattered, more real say."[34] Williams will not gainsay these freedoms; he and his family, and all those others who occupy ordinary culture, will not give up this power for anything.

There are vast differences in energy use across the world, and within different groups and communities within every nation. Many parts of the world are still awaiting the capacities and opportunities that Williams's home experienced in 1958. The International Energy Association estimated total world energy consumption in 2013 was 12.3 terawatts (a terawatt is one trillion watts). This energy is used unequally, ranging from 13,000 to 25,000 W per capita in places like Kuwait and Qatar to 286 W in Bangladesh and 390 W in the Congo—a *one-hundred-fold* difference in energy use between Qatar and Bangladesh, and even more if compared to Afghanistan or Eritrea.[35] These differences between countries speak to differences of climatic zone, the size of countries, and more. But they also speak to vast differences of social and physical capacity and ability granted to people around the world as a result of the energy to which they have had access. Put bluntly: in many parts of the world women still have to wash clothes by hand in (often) dirty rivers, and children have to walk miles to and from school. This is, at a minimum, time and physical and mental energy that could be put to different use.

In the world today, not everyone has the basic levels of energy that Williams insists on as necessary for freedom. Those who have yet to realize this level of energy are not ready to simply give up on the capacities and possibilities that increasing levels of fossil-fuel use opens up simply because those in the Global North have suddenly become anxious about the deformation of atmospheric gases that their societies have generated over two centuries of economic value extraction from the environment. This presents an enormous political challenge, one that I believe will come to define the politics of this century. The problem of not recognizing the role of cheap energy in freedom and democracy is that we've avoided taking on the material challenge of reimagining opportunities for more and more people around the globe, preferring the story of a progressivist narrative whereby things slowly get better with time, no matter what. The unspoken hope and expectation is that the developing world will leapfrog the fossil-fuel era straight into solar panels and wind farms, and thus avoid linking development with the burning of carbon. How exactly this is supposed to occur is never fully articulated or explained—it's just supposed to happen so that we might be saved from the trouble of attending to the deep environmental and economic injustices that access to dirty energy has fueled and continues to fuel.

How else might we approach these issues of justice and injustice in relation to energy and the freedoms that accompany its use? In *A Theory of Justice* (1971), political philosopher John Rawls famously begins his elaboration of the principles of social justice by articulating a thought experiment—the "original position," which is a hypothetical ground zero from which the principles of liberal societies were re-constituted. How might we develop social justice today were we to start from scratch, shrouded behind a veil of ignorance, unaware of what position we might occupy in society, or our ethnicity or gender, or anything at all? What principles of justice might we establish that would value each of us for who and what we are *qua* being human? Much of Rawls's lasting influence on discussions of ethics and politics within the liberal tradition is owed to the conceptual power of this original position, which asks us to consider the levels of equity, fairness, and justice that should be extended to everyone.

What if we were to add energy to the issues that had to be addressed in this original position? Rawls never speaks of energy as an issue of social justice. And while he is not a strict egalitarian in his understanding of how goods, capacities, and abilities should be assigned in a society, he does identify the need for there to exist a social minimum available to each and every person such that they can achieve their version of the good life however they might want. In the assignment of how much energy each person should have available to them, it's unlikely that those in the original position tasked with creating the principles of social justice would think it fair or reasonable that there be vast differences in the amount of energy available to each person: it would mean vast differences in the capacities and opportunities for individuals; it would also mean huge differences in environmental impact across communities, with those using very little energy having to live in an atmosphere poisoned by those using a great deal—in other words, a model of the world as it presently exists.

The average per capita energy use across the globe is 1640 W. Might this represent the beginning point of a discussion over what an energy commons might look like? Of what might constitute a just use of energy? If everyone on the planet used the same energy as a Canadian, total planetary consumption would be 74 terawatts per year—*six times* as much energy as we currently consume across the globe. The figure of 1640 W per capita is close to current energy use in places like Uruguay and Iraq; and even this figure is too high given the need to limit energy even further due to increases of population and, of course, impact on the environment.

The political challenge posed by energy justice is enormous, involving not just the generation of new social narratives, but also the production of new infrastructures. What is required "is a radical change in the key economic choices which shape civilization over long periods . . . a decisive broadening of political and social democracy, a profound change in individual behavior and education."[36] This is why energy transition necessitates a social, cultural, and political transition, too, a profound re-imagining of resource demands and their environmental consequences. What is needed for a real energy transition is something akin to a revolution, if for reasons and on terms very different from how revolution was understood over the previous two centuries. A critical theory of energy *is* about our collective emancipation from slavery; it is a politics of freedom, democracy, and equity that always already attends to the resources that expand and extend human capacity, and ensures that our use of these resources doesn't impact the physical environments we collectively inhabit.

I hope that the insights these essays bring together about the difficult issues and questions we face, involving the goods of the Earth and the goods we expect to have in our lives, might contribute to the ongoing struggle to bring social and environmental justice together in rich and vibrant new ways. I hope, too, that they help set the terms on which it might be possible to talk of justice and revolution not only with my academic peers, but also with acquaintances in dog parks or cafes—with all those people whose ordinary cultures will need to be honored and energized in the many transitions just over the horizon.

—Toronto, February 2018

NOTES

1. Canada remains far off the climate target it set for itself under the terms of the 2015 Paris Agreement. Emissions in 2015 were 722 megatonnes of greenhouse gas emissions; the 2030 target is 523 megatonnes. See Marie-Danielle Smith, "Emissions down slightly, but Canada not yet on track to meet 2030 climate targets: report," *National Post*, Apr. 21, 2017, http://nationalpost.com/news/politics

/emissions-down-slightly-but-canada-not-yet-on-track-to-meet-2030-climate-targets-report (accessed August 17, 2018).

2.   Where did my coffee shop pal get this info? I expected to be able to find some slip of evidence online on various climate denial websites. In this case, to my surprise, I simply wasn't able to do so.

3.   A report released by the Arctic Council in November 2017, which summarized research from 2010 to 2016 and represented the work of 90 scientists, concludes "the top of the world is getting warmer faster than anyone thought . . . the Arctic continues to warm at twice the pace of mid-latitudes and is likely to see warming of up to five degrees Celsius as early as 2040." Bob Weber, "Arctic warming happening faster than previously thought, report says," *Globe and Mail*, Nov. 19, 2017, https://www.theglobeandmail.com/news/national/arctic-warming-happening-faster-than-previously-thought-report-says/article 37024293/ (accessed August 17, 2018).

      2017 was the hottest recorded year in the Earth's oceans and the second hottest at Earth's surface. See John Abraham, "In 2017, the oceans were by far the hottest ever recorded," Guardian, Jan. 26, 2016, https://www.theguardian.com/environment/climate-consensus-97-per-cent/2018/jan/26/in-2017-the-oceans-were-by-far-the-hottest-ever-recorded?CMP=Share_iOSApp_Other (accessed August 17, 2018).

4.   See Naomi Orestes' now classic *Science* essay "The Scientific Consensus on Climate Change." More than a decade later, the strong scientific consensus on anthropogenic global climate change noted by Orestes has become even stronger.

5.   I'm pretty certain where my co-caffeine liquid got his info about *An Inconvenient Truth*. The fourth item in a Google search of "Inconvenient Truth 10 points" is a page called "8 Highly Inconvenient Facts for Al Gore 10 Years After His Infamous Movie." See http://www.theblaze.com/contributions/8-highly-inconvenient-facts-for-al-gore-10-years-after-his-infamous-movie (accessed August 17, 2018). In the third item in the search, "What 'An Inconvenient Truth' Got Right (And Wrong) About Climate Change," Patrick J. Kiger asks scientists to check in with Gore's movie to see where things stand a decade later. While there might be some things off about *An Inconvenient Truth*, in this same article Ted Scambos, a senior scientist at the US National Snow and Ice Data Center, points out "the basic truth, and its inconvenience, remains. In fact, it is clearer than ever that greenhouse gases are a major cause of the observed climate warming." On the Internet, it seems, one can find "evidence" for whatever position one has adopted.

6.   For an analysis of the challenges of communicating about climate change to oft-skeptical communities, see Callison, *How Climate Change Comes to Matter*.

7.   The difficulty of communicating effectively about environmental issues is at the center of the field of environmental communication. See, for example, Hansen, *Environment, Media and Communication*, and Hansen and Cox, eds., *The Routledge Handbook of Environment and Communication*.

8.   Indeed, critics have suggested that this is at the heart of anxieties linked to the consequences of climate change. For instance, Ursula Heise argues that "biodiversity, endangered species, and extinction are primarily cultural issues, questions of what

we value and what stories we tell, and only secondarily issues of science." Heise, *Imagining Extinction*, 5.

9. Up-to-date information on the environment produced by the UN's Intergovernmental Panel on Climate Change can be found online at http://www.ipcc .ch, while the text of the Paris Agreement can be located at: http://www.cop21.gouv .fr/en (accessed August 17, 2018). A definitive pronouncement of the causes and consequences of climate by the Union of Concerned Scientists has been posted at: https://www.ucsusa.org/global-warming/science-and-impacts/science/scientists -agree-global-warming-happening-humans-primary-cause#.Wik06yOZM1i (accessed August 17, 2018). Finally, see Pope Francis, *Encyclical on Climate Change and Inequality*.

10. Barthes describes right-wing myth as "well-fed, sleek, expansive, garrulous, it invents itself ceaselessly. It takes hold of everything, all aspects of the law, of morality, of aesthetics, of diplomacy, of household equipment, of Literature, of entertainment." Barthes, *Mythologies*, 148.

11. For Timothy Morton, true environmental thinking has been blocked by approaches to the environment that re-inscribe the instrumentality to which Martin Heidegger drew our attention. What Morton terms "hyperobjects" are objects, events, and experiences that exceed the individual and local, and so cannot be meaningfully reigned in through existing forms of systemic thought. One of the most important hyperobjects for Morton is the global climate system; another is oil and energy. If the latter has escaped extant epistemologies, and the former exceeds any ability to conceptually frame it (so that, for instance, one might know what "it" is doing in any given place at any given time), it is because the scale of both is such that they necessarily exist outside of existing systems of thought. Morton, *Hyperobjects*.

12. Szeman, *Zones of Instability*.

13. "Every nation is one people," Herder writes, "having its own national form, as well as its own language." Herder, *Outlines of the History of Man*.

14. Bauman, *Globalization*, 77–102.

15. Ghosh, "Petrofiction: The Oil Encounter and the Novel."

16. For an overview of the field of energy humanities, see Szeman and Boyer, eds., *Energy Humanities*, and Macdonald and Stewart, eds., *Routledge Handbook of Energy Humanities*. A text that promises to take the field in exciting and intriguing new directions is Bellamy and Diamanti, *Materialism and the Critique of Energy*, a collection that includes contributions from thinkers such as George Caffentzis, Andreas Malm, and Alberto Toscano, among others.

17. Ghosh, "Petrofiction" and *The Great Derangement*.

18. Stoekl, *Bataille's Peak*; Chakrabarty, "The Climate of History: Four Theses"; Mitchell, *Carbon Democracy*; Diamanti, "Energyscapes, Architecture, and the Expanded Field of Postindustrial Philosophy" and "Three Theses on Energy and Capital"; LeMenager, "Petro-Melancholia: The BP Blowout and the Arts of Grief" and *Living Oil*; and Wenzel, "Introduction" to *Fueling Culture: 101 Words for Energy and Environment*.

19. Horkheimer, *Critical Theory*, 246. Marx's eleventh thesis reads: "The philosophers have only *interpreted* the world, in various ways; the point, however, is to *change* it."

20. Christopher Jones has accused energy humanities scholars of displaying a

reductive, "petromyopia," one in which it becomes possible, in the end, to explain everything as constituted by fossil fuels. As my comments here should suggest, this is to miss the point of the critical intervention made by energy humanities. Jones, "Petromyopia: Oil and the Energy Humanities," n.p.

21. Mouhout, "Past Connections and Present Similarities in Slave Ownership and Fossil Fuel Usage," 339–40. For other such connections between energy and slavery, see Debeir, Deléage, and Hémery, *In the Servitude of Power*; Johnson, "Energy Slaves: The Technological Imaginary of the Fossil Economy" in *Mineral Rites*; and Nikiforuk, *The Energy of Slaves*.

22. McNeill and Engelke, *The Great Acceleration*.

23. McNeill provides a full accounting of the massive expansion in resource use, especially since World War II, in McNeill, *Something New Under the Sun*.

24. See German Advisory Council on Global Change, *World in Transition 3: Towards Sustainable Energy Systems*; German Federal Ministry for Economic Affairs, *The Energy of the Future*; and Morris and Jungjohann, *Energy Democracy*.

25. With respect to energy, at least, Germany might be an exception: "The Energiewende is nonetheless exceptional in one way too often overlooked. Apart from Denmark and, more recently, Scotland, Germany is the only country in the world where the switch to renewables is a switch to energy democracy." Morris and Jungjohann, *Energy Democracy*, x.

26. It is unlikely that capitalism could continue easily on, just powered in a different fashion. One main reason is due to the significant changes in Energy Return on Energy Invested (EROEI) over the life of oil capitalism. Current figures suggest that EROEI has been as high as 45 in the 1960s, but has declined to an average of 20, and continues to decline. EROEI on photovoltaics is around 10; for wind the figure is about 20—both well below the cheap energy around which modernity was constituted. See Murphy, "The Implications of the Declining Energy Return on Investment of Oil Production."

   As Jason Moore points out, crises of capitalist accumulation have frequently been managed by appropriating cheap nature. The effective end of cheap nature suggests that such a maneuver is fast approaching an end. See Moore, *Capitalism in the Web of Life*.

27. Even if this were possible—a big if—we would still have to engage in extractivist practices that *cannot* be imagined in relation to a transition in quite the same manner as energy (i.e., one cannot switch to renewable iron ore or copper). Energy use and practices of extraction are closely linked. See the 2017 special issue of *Cultural Studies* on "Cultural Studies of Extraction," edited by Laura Junka-Aikio and Catalina Cortes-Severio, which includes my afterword, "On the Politics of Extraction," 440–47.

28. Chakrabarty, "The Climate of History: Four Theses." Timothy Mitchell puts it just as directly: "Fossil fuels helped create both the possibility of twentieth-century democracy and its limits." Mitchell, "Carbon Democracy" (2009), 399.

29. Ian Baucom has written: "Although I have for some time accepted the force of Fredric Jameson's dictum that 'we cannot, not periodize,' until very recently it would not have occurred to me that postcolonial study, critical theory, or the humanities disciplines in general needed to periodize in relation not only to capital

but to carbon, not only in modernities and post-modernities but in parts-per-million, not only in dates but in degrees Celsius." Baucom, "History 4°: Postcolonial Method and Anthropocene Time," 125.

30. In "Exploring the anthropology of energy: Ethnography, energy, and ethics," Jessica Smith and Mette High argue that it is necessary to attend to energy ethics in considering energy justice. They write: "We emphasize that our approach takes seriously people's own ethical sensibilities in relation to energy, working from the ground up, rather than analyzing social life through pre-defined notions of ethics. Energy ethics illuminates the multiple and varied ways that people experience, conceptualize, and evaluate matters of energy in their lives" (n.p.). This attentiveness to the specificity of the distinct uses of energy by communities around the globe is crucial, both as a guiding protocol for the practice of energy anthropology and to give greater nuance to our understanding of energy justice. A motto for energy justice might be "From each according to their energy capacities, to each according to their energy needs (within environmental limits)." See Smith and High, "Exploring the anthropology of energy," 1.

31. Arnold, *Culture and Anarchy*, 7.

32. Williams, "Culture Is Ordinary," 57.

33. Williams, "Culture Is Ordinary," 56.

34. Williams, "Culture Is Ordinary," 56.

35. All figures on energy use are from World Bank, *World Development Indicators*. Available at: https://datacatalog.worldbank.org/dataset/world-development-indicators

36. Debeir, Deléage, and Hémery, *In the Servitude of Power*, 237.

CHAPTER 1

Who's Afraid of National Allegory? Jameson, Literary Criticism, Globalization (2001)

Fredric Jameson's proposal that all third-world texts be read as "national allegories" has been one of the more influential and important attempts to theorize the relationship of literary production to the nation and to politics. Unfortunately, its influence and importance has thus far been primarily *negative*. For many critics, Jameson's essay stands as an example of what not to do when studying third-world literature from the vantage point of the first-world academy. His attempt in the now infamous essay, "Third-World Literature in the Era of Multinational Capitalism," to delineate "some general theory of what is often called third-world literature" has been attacked for its very desire for generality.[1] The presumption that it is possible to produce a theory that would explain African, Asian, *and* Latin American literary production, the literature of China *and* Senegal, has been (inevitably) read as nothing more than a patronizing, theoretical orientalism, or as yet another example of a troubling appropriation of Otherness with the aim of exploring the West rather than the Other. The most well-known criticism of Jameson's essay along these lines remains Aijaz Ahmad's "Jameson's Rhetoric of Otherness and the 'National Allegory'."[2] More informally and anecdotally, however, within the field of postcolonial literary and cultural studies, Jameson's essay has come to be treated as little more than a cautionary tale about the extent and depth of Eurocentrism in the Western academy, or, even more commonly, as a convenient bibliographic marker of those kinds of theories of third-world literature that everyone now agrees are limiting and reductive.[3]

Looking back on Jameson's essay through the haze of fifteen years of postcolonial studies, as well as through the equally disorienting smoke thrown up by the explosion of theories and positions on globalization, one wonders what all the fuss was about. In hindsight, it appears that almost without exception critics of Jameson's essay have willfully misread it. Of course, such misreadings are to be expected. The reception given to this or that theory has as much to do with timing as with its putative content. As one of the first responses to postcolonial literary studies from a major critic outside the field, the publication of Jameson's essay in the mid-1980s provided postcolonial critics with a flash point around which to articulate general criticisms of dominant views of North-South relations expressed within even supposedly critical political theories (like Marxism). It also provided a self-definitional opportunity for postcolonial studies: a shift away from even the lingering traces of Marxist interpretations of imperialism toward a more deconstructive one exemplified by the work of figures such as Gayatri Chakravorty Spivak and Homi Bhabha.[4] While criticisms of Jameson's views may have thus been useful or productive in their own way, they have nevertheless tended to obscure and misconstrue a sophisticated attempt to make sense of the relationship of literature to politics in the decolonizing world. I want to argue here that Jameson's "general" theory of third-world literary production offers a way of conceptualizing the relationship of literature to politics (and politics to literature) that goes beyond the most common (and commonsense) understanding of the relations between these terms.[5] Indeed, the concept of national allegory introduces a model for a properly materialist approach to postcolonial texts and contexts, one that resonates with Kalpana Seshadri-Crooks's recent characterization of postcolonial studies as "interested above all in materialist critiques of power and how that power or ideology seems to interpellate subjects within a discourse as subordinate and without agency."[6]

Seshadri-Crooks's description of the aims of postcolonial studies emerges out of her analysis of the malaise or melancholia that has beset postcolonial studies as it enters the new millennium. It seems to me that revisiting Jameson's theory of third-world literature—both its problems and its productive potentialities—provides a (perhaps unexpected) way out of this malaise. One of the things for which Jameson has been criticized throughout his career is his insistence on totality as a central concept in social and political criticism. In the context of postcolonial studies, it is easy enough to see how this appeal to totality could be mistaken as a Eurocentric, universalist claim par excellence.[7] But this is to conceive of the concept of totality far too rigidly and unimaginatively, and in the process of doing so, to "fall back into a view of present history as sheer heterogeneity, random difference, a coexistence of a host of distinct forces whose effectivity is undecidable."[8] It seems to me that what is missing in most theories of postcolonial literary production (and what thus produces the malaise that Seshadri-Crooks points to) is just such a map of the relative effectivity of different forces in the globalscape: in the absence of some general theory of the structure of contemporary social and political life, there is instead a rough assemblage of literary-critical commonplaces concerning power, identity, representation, language, and so on, that originate almost solely from within the hermetically sealed space of academic criticism.[9] In any case, my argument here should also be taken as an implicit argument on behalf of totality—not the "bad" totality that legitimates theories of modernization of development, but the totality constructed by an antitranscendental and antiteleological "insurgent science" that "is open, as open as the world of possibility, the world of potential."[10] Here, at least, totality appears as the possibility of metacommentary—not as a secondary step in interpretation but as a condition of interpretation per se; and as I argue here, what national allegory itself names are the conditions of possibility of metacommentary at the present time.[11] The question I pursue, then, is the relationship of allegory (as a mode of interpretation) to the nation (as a specific kind of sociopolitical problematic) and what this relationship entails for a global or transnational literary or cultural criticism.

In an effort to uncover the possibilities and limits of the concept of national allegory, I first reexamine Jameson's development of the concept of national allegory in "Third-World Literature in the Era of Multinational Capitalism." I then turn to a consideration of the history of this term in Jameson's own work. While it has been stressed that Jameson's comments concerning third-world literature arise out of meditations on a different matter entirely (that is, the debates in the American academy in the mid-1980s over the revision of the literary canon), almost no critic has made reference to the fact that the concept of national allegory does not originate in this essay.[12] Finally, I ponder the relationship between nation and allegory by looking at Jameson's recent writings on globalization in order to consider its significance for contemporary cultural theory and criticism.

One of the first things that has to be made clear about Jameson's account of third-world literature is that the concept of national allegory is exhausted by neither of its component terms. Jameson is aware that the *nation* and *allegory* are concepts that have both fallen into disrepute: the nation, because of the historical experiences of first- and third-world countries with the virulent nationalisms of the twentieth century, as well as the vigorous criticism that has been directed toward the nation over the past several decades; allegory, because of the naive mode of one-to-one mapping that it seems to imply, a presumed passage from text to context that is epistemologically and politically suspect. Attaching these terms to a theory of third-world texts has a tendency to conjure up once again the whole specter of development theory and practice, in which technologies that have become antiquated in the West are passed along to countries where such outmoded technologies (including conceptual technologies such as the nation and allegory) might, in Hegelian fashion, still be of some use. There is no doubt that some of the initial discomfort felt by many critics with the concept of national allegory arises out of a resistance to the political implications of each of its component terms—to the sense, that is, that either of these terms still has a relevance for the "underdeveloped" third world that they have (as Jameson admits) lost in the "developed" first (in this way becoming the literary-critical equivalent of pesticides long banned in the West that continue to be produced in the United States for sale in the third world).

Infamously, Jameson writes that "all third-world texts are necessarily . . . allegorical, and in a very specific way they are to be read as what I will call *national allegories*."[13] Here again, the claim that Jameson makes about third-world texts ("by way of a sweeping hypothesis") cannot help but distract from his broader aim, which is not to pass aesthetic judgment on third-world texts, but to develop a system by which it might be possible to consider these texts *within* the global economic and political system that produces the third world as the third world.[14] For Jameson, third-world texts are to be understood as national allegories specifically in contrast to the situation of first-world cultural and literary texts. He argues that there is a political dimension to third-world texts that is now (and has perhaps long been) absent in their first-world counterparts. This corresponds to a difference between the social and political culture of the first and third worlds—a difference that must, of course, be understood as broad and conceptual, and that should not be seen as unreflectively rendering homogenous what are two extraordinarily heterogeneous categories.[15] Jameson believes that in the West, the consequence of the radical separation between the public and the private, "between the poetic and the political," is "the deep cultural conviction that the lived experience of our private existences is somehow incommensurable with the abstractions of economic science and political dynamics."[16] In terms of literary production, this "cultural conviction" has the effect of limiting or even negating entirely the political work of literature: in the first world, literature is a matter of the private rather than the public sphere, a matter of individual tastes and solitary meditations rather than public debate and deliberation. The relations between the public and the private in the third world are entirely different: they have not undergone this separation and division. Literary texts are thus never *simply* about private matters (although, as Michael Sprinker points out in his review of Jameson's essay, they are never *simply* private in the first world either, however difficult it might be to see this now).[17] In the third world, Jameson claims, "the story of the private individual destiny is always an allegory of the embattled situation of the public third-world culture and society."[18]

This claim is strong and sweeping, one whose precise meaning in "Third-World Literature in the Age of Multinational Capitalism" can be grasped only by careful attention to Jameson's description of allegory, his claims about the relationship of psychology to politics in the first and third worlds, and his description of the significance of the term *culture* and the relationship between culture and politics more generally. Of the concept of allegory, Jameson writes that "our traditional concept of allegory—based, for instance, on stereotypes of Bunyan—is that of an elaborate set of figures and personifications to be read against some one-to-one table of equivalence: this is, so to speak, a one-dimensional view of this signifying process, which might only be set in motion and complexified were we willing to entertain the more alarming notion that such equivalencies are themselves in constant change and transformation at each perpetual present of the text."[19] Read in this more expansive way, the allegorical mode is not limited to the production of morality tales about public, political events—tales that could just as well be described in journalistic terms as in the narrative structure of novels or short stories. On the contrary, "the allegorical spirit is profoundly discontinuous, a matter of breaks and heterogeneities, of the multiple polysemia of the dream rather than the homogenous representation of the symbol."[20] If in the third world, private stories are always allegories of public situations, this does not thereby imply that of necessity third-world writing is narratively simplistic or overtly moralistic, or that all such texts are nothing more than exotic versions of Bunyan, as might be supposed in the terms of a more traditional sense of allegory. The claim is rather that the text speaks to its context in a way that is more than simply an example of Western texts' familiar "auto-referentiality": it necessarily and directly speaks to and of the overdetermined situation of the struggles for national independence and cultural autonomy in the context of imperialism and its aftermath.[21]

Why third-world texts speak more directly of and to the national situation has to do with what Jameson sees as the very different "relationship between the libidinal and the political components of individual and social experience" in the first and third worlds.[22] One of the results of the deep division between the public and private spheres in the first world is that "political commitment is recontained and psychologized or subjectivized."[23] Again, for Jameson, the very opposite is the case in the third world. The division between public and private that is characteristic of the West is not characteristic of most third-world societies, or perhaps this should be read as "not yet" or "not yet completely."[24] This assertion could be taken (again, in Hegelian fashion) as a claim that, socially and aesthetically, the third world lags behind the first in its development.[25] But—and I think this is how Jameson intends it—it also highlights a genuine, material difference between the first and third worlds that is expressed socially and culturally. The attempt to maintain a different form of social life while accepting the material and technological advantages offered by the West has constituted one of the major challenges faced by non-Western societies for whom modernity has been belated; it does not seem to me inconceivable to imagine a different organization of private and public in societies that were the subjects of colonialism as opposed to its agents.[26] In any case, the lack of a corresponding division between public and private in the third world means for Jameson that "psychology, or more specifically, libidinal investment, is to be read in primarily political and social terms."[27] If political energies in the first world are psychologically interiorized in a way that divests them of their power, it could be said that in the third world the "sphere" of the psychological does not function as a containment device in which what is dangerous in the public is sublimated and defused. In the first world, these sublimated energies may, of course, return to the public sphere in the mediated form of various cultural products; even so, unlike the situation of the third world, in the first world such cultural products would nevertheless be taken to be imbued with only *private* significance or with only the most banal form of larger public meaning, that is, as indicators of "styles" or "trends," the Hegelian *Geist* reborn as successive waves of (essentially similar) commodities. Another way of characterizing this division between first and third worlds within Jameson's own vocabulary is to say that the history that is everywhere actively repressed in the first world is still a possible subject of discourse in the third world (consider, for instance, his discussion of the repressed spaces of Empire in British modernism).[28] Of course, this characterization of the large-scale societal differences between the first and third worlds, Jameson adds, must be read as "speculative" and general, and open to "correction by specialists."[29]

Jameson's characterization of the different relationships in the first and third worlds between private and public, and so also of the psychological or the libidinal, must be read further in terms of his subsequent discussion of the concept of *cultural revolution*; otherwise, it is possible at this point to see his characterization of the vast social, political, and cultural gulf separating the first and the third worlds as a form of Eurocentrism or exoticism in which—as in the early moments of modernist art—what is lacking in the civilized West is found at the heart of its "uncivilized" exterior. Jameson links the idea of "cultural revolution," which has most commonly been used to refer to the massive set of social and cultural changes undertaken by communist regimes (and in China in particular), to the work of figures with "seemingly very different preoccupations": Antonio Gramsci, Wilhelm Reich, Frantz Fanon, Herbert Marcuse, Rodolph Bahro, and Paolo Freire. It is in the connection that Jameson makes between cultural revolution and "subalternity" that the significance of "national allegory" as an interpretive strategy for third-world texts begins to come into focus:

> Overhastily, I will suggest that "cultural revolution" as it is projected in such works [Gramsci, Reich, et al.] turns on the phenomenon of what Gramsci called "subalternity," namely the feelings of mental inferiority and habits of subservience and obedience which necessarily and structurally develop in situations of domination—most dramatically in the experience of colonized peoples. But here, as so often, the subjectivizing and psychologizing habits of first-world peoples such as ourselves can play us false and lead us into misunderstandings. Subalternity is not in that sense a psychological matter, although it governs psychologies; and I suppose that the strategic choice of the term "cultural" aims precisely at restructuring that view of the problem and projecting it outwards into the realm of objective or collective spirit in some non-psychological, but also non-reductionist or non-economistic materialist fashion. When a psychic structure is objectively determined by economic and political relationships, it cannot be dealt with by means of purely psychological therapies; yet it equally cannot be dealt with by means of purely objective transformations of the economic and political situation itself, since the habits remain and exercise a baleful and crippling residual effect. This is a more dramatic form of that old mystery, the unity of theory and practice; and it is specifically in the context of this problem of cultural revolution (now so strange and alien to us) that the achievements and failures of

third-world intellectuals, writers and artists must be placed if their concrete meaning is to be grasped.[30]

So the concept of national allegory highlights the ways in which the psychological points to the political and the trauma of subalternity finds itself "projected outwards" (allegorically) into the "cultural." Very crudely, the cultural is what lies "between" the psychological and the political, unifying "theory and practice" in such a way that it is only there that the "baleful and crippling" habits that are the residue of colonialism can be addressed and potentially overcome. A "cultural revolution" aims to do just this—to produce an authentic and sovereign subjectivity and collectivity by undoing the set of habits called subalternity. While these are not habits that can be modified by the transformation of political and economic institutions alone, this does not mean the exclusive attention to the subjective (the psychological) or to the cultural is sufficient in and of itself either. The idea of *habit* is for this reason a particularly apt way of understanding the legacy of subalternity, since it draws attention to the ways in which subalternity cannot be reduced simply to "mental" or "psychological" states, but must be seen as residing in the unconscious and inscribed somatically in a whole range of bodily dispositions. The problem of cultural revolution accounts for the presence of the political in the psychological by means of a level of mediation comprised of cultural objects like literary texts, and provides a framework in which it is possible to assess "the achievements and failures of third-world intellectuals" with respect to the task of reclaiming something positive from the colonial experience.[31]

The relationship between the cultural and subalternity may be seen, of course, as almost generically definitive of the intellectual work that has been produced under the sign of "postcolonial" theory and criticism. For example, to point to one of the earliest works (retrospectively) in postcolonial criticism, what other than the "habit" of subalternity does Frantz Fanon address in *Black Skin, White Masks*? One of the most important things that postcolonial critics have added to our understanding is the degree to which cultural and discursive domination was (and is) a necessary and essential aspect of colonialism and imperialism.

Where Jameson differs from most postcolonial critics, however, is in his insistence that "culture"

> is by no means the final term at which one stops. One must imagine such cultural structures and attitudes as having been themselves, in the beginning, vital responses to infrastructural realities (economic and geographic, for example), as attempts to resolve more fundamental contradictions—attempts which then outlive the situations for which they were devised, and survive, in reified forms, as "cultural patterns." Those patterns themselves then become part of the objective situation confronted by later generations.[32]

He continues,

> Nor can I feel that the concept of cultural "identity" or even national "identity" is adequate. One cannot acknowledge the justice of the general poststructuralist assault on the so-called "centered subject," the old unified ego of bourgeois individualism, and then resuscitate this same ideological mirage of psychic unification on the collective level in the form of a doctrine of collective identity. Appeals to collective identity need to be evaluated from a historical perspective, rather than from the standpoint of some dogmatic and placeless "ideological analysis." When a third-world writer invokes this (to us) ideological value, we need to examine the concrete historical situation closely in order to determine the political consequences of the strategic use of this concept.[33]

There are then (at least) two levels of mediation that must be considered in the movement from the psychological to the political (and back again) through the cultural. Culture mediates; to understand precisely how it does so, it must be understood that the cultural forms and patterns that produce this mediation are themselves the product of an earlier process of mediation—now reified into the forms and patterns of culture that are to be used as the raw materials of cultural production. Few critics now would object to the need for the analysis of any form of cultural production to take into account the circuits of economics and politics that make the text possible in the first place, but the significance of this second mode by which culture mediates remains all too often unexplored. In other words, what is often missing is the realization that all mediation in the present takes place through the reified cultural forms (and culture in general) of the past; all attempts to resolve the "fundamental contradictions" of the present through cultural production must pass through the concretized history of previous attempts to solve the contradictions of earlier infrastructural realities that have since changed in form and character. This is not to say that culture must be understood as somehow necessarily belated, or that it therefore always "misses" the present, which is to misunderstand in any case what it might mean for cultural forms to attempt to resolve historical contradictions. It is, rather, to point out the need for a more complicated understanding of the process of mediation that considers not simply the site of mediation (say, the text), but also the way in which this site is itself the product of mediation. It is this sense of mediation to which Adorno was trying to draw our attention, too, when he said "mediation is in the object itself not a relation between the object and those to whom it is brought."[34]

Far from reducing the complexity of third-world literary production, the concept of national allegory enables us to consider these texts as the extremely complex objects that they are and not just as allegories of one kind or another of the Manichean binaries produced out of the encounter of colonizer and colonized (however ambivalently one might want to understand these). It also foregrounds (metacritically) the cultural/social situation of the reader of the texts, and indeed, the very fact that every interpretation or reading is a kind of translation mechanism that it is best to acknowledge rather than to hide the workings of; the critic, too, works out of a cultural situation that is the product of earlier mediations that form the raw material for his or her readings.[35] Understood through the lens of the idea of cultural revolution that Jameson outlines here, the concept of national allegory suggests a number of things about how we should think about postcolonial or third-world texts in the context of the period of decolonization and globalization. First, postcolonial literary production needs to be understood as forming "vital responses to infrastructural realities . . . as attempts to resolve more fundamental contradictions." In other words, it is productive to look at this form of cultural production as a particular kind of cultural strategy, rather than "read" simply and immediately as "literature," in the sense in which this concept is well understood in the first-world academy.[36] Second, careful attention needs to be paid to the deployment of "ideological values" by third-world writers themselves, values that sometimes have a resonance in the Western academy because of the ways in which they politically re-empower the project of Western literary criticism. One of the most important of these may be that of the "nation" and its strategic use in the literature produced during decolonization; another is to be found in the unquestioned assumption on the part of many critics of the almost necessary social significance of postcolonial literature (or at least, its significance in a straightforward way), when literature may in fact have a relatively marginal role in the postcolony. Another way of putting this last point is that in the examination of postcolonial literature, what needs to be considered are the conditions of possibility for the practice of writing literature in these regions, for it is only in this way that we can understand the precise and complicated ways in which this older, imported "technology" participates in the task of cultural revolution that is so important to third-world societies.[37]

Whatever one might think of this formulation of mediation and of its utility for postcolonial literary studies, it might nevertheless seem as if I have come far afield from the initial concept of national allegory in producing it. This elaboration of national allegory appears to be more or less akin to the general interpretive schema that Jameson has developed with remarkable consistency over the course of his career, specifically in works such as *The Political Unconscious*. And if *this* is what national allegory is finally about, one has to wonder why Jameson would have generated a neologism that cannot help but invite confusion. Why, after all, *national* allegory and not something else? In elaborating how this mode of interpretation has specific relevance to the theorization of the role and function of culture and literature in the era of globalization, I want to briefly review the history of national allegory in Jameson's own work. For if there is anything that is troubling about the use of national allegory as a mode of analysis of third-world literary texts, it is to be found in the changes that this concept undergoes throughout

Jameson's work, coming to be, finally, nothing less than a substitute term for the kind of dialectical criticism that he would like to apply to all cultural texts—whether third world or not.[38] National allegory names a possibility and a limit for texts that Jameson first sees in the fiction of Wyndham Lewis, then in third-world texts, and finally, as a condition of contemporary cultural production as such. What is missing in Jameson's discussion of national allegory is a discussion of the *nation* to match that of *allegory*. Though it might seem as if the nation has an important role to play in understanding third-world texts, on the question of the nation itself, Jameson has surprisingly little to say in "Third-World Literature": the nation is more or less simply conflated with the "political" and, when it is not, it becomes a term that seems to make reference to a kind of collectivity or community that is idealized when it should be placed into question. It is in this lack of attention to the issue of the nation in the concept of national allegory that the strains of the transposition of this concept from an earlier formulation become apparent. While there are thus limits to national allegory within "Third-World Literature," it seems to me that looking at some of Jameson's more recent reflections on the nation in the context of globalization can help to locate the nation within his dialectical mode of analysis in a way that brings national allegory forward into the global present even as it clarifies the conceptual work that the nation performs in his analysis of third-world texts.

The term "national allegory" first appears in *Fables of Aggression* as a description of Wyndham Lewis's novel, *Tarr*. As it is presented in this early work, national allegory originates as a much more straightforward concept than it comes to be in the discussion of third-world texts: it refers to the way in which individual characters with different national origins stand in for "more abstract national characteristics which are read as their inner essence."[39] When dealing with any one such correspondence between character and national essence, this allegorical mode becomes a form of "cultural critique." For Jameson, the unique characteristic of Lewis's texts is to have assembled numerous national types into one setting, thereby producing "a dialectically new and more complicated allegorical system . . . that specific and uniquely allegorical space between signifier and signified."[40] In *Fables of Aggression*, national allegory is thus the name for a specific, formal characteristic of Lewis's novel, rather than a concept that suggests an entire system or mode of reading and interpretation. Indeed, the more general logic that Jameson suggests as the only way to properly account for the possibility in Lewis's novel of this "now outmoded narrative system" seems to have become transformed with reference to third-world texts into the principle of what is now national allegory itself.[41] In characteristic form, Jameson draws attention to the fact that an explanation for national allegory as a formal principle of *Tarr* can only be found in history—though not in the sense that historical conditions "caused" the formal organization of *Tarr* or that the novel is "a 'reflexion' of the European diplomatic system."[42] Instead, he suggests, our attention should be directed toward

> the more sensible procedure of exploring those semantic and structural givens which are logically prior to this text and without which its emergence is inconceivable. This is of course the sense in which national allegory in general, and *Tarr* in particular, presuppose not merely the nation-state itself as the basic functional unit of world politics, but also the objective existence of a system of nation-states, the international diplomatic machinery of pre-World-War-I Europe which, originating in the 16th century, was dislocated in significant ways by the War and the Soviet Revolution.[43]

According to Jameson, all literary and cultural forms provide an "unstable and provisory solution to an aesthetic dilemma which is itself the manifestation of a social and historical contradiction."[44] National allegory can therefore be seen as a once but no longer viable formal attempt "to bridge the increasing gap between the existential data of everyday life within a given nation-state and the structural tendency of monopoly capital to develop on a world-wide, essentially transnational scale."[45] In other words, the formal qualities of *Tarr* point to the fact that life in England can no longer be rendered intelligible with the "raw materials" of English life alone; narrative resources must be sought elsewhere, and what lies "outside" England is for Lewis (objectively and structurally) a system of nation-states (and their attendant national cultures): "The lived experience of the British situation is domestic, while its structural intelligibility is international."[46]

It is striking that the words Jameson uses to describe the "problem" to which Lewis's national allegory is a solution are almost exactly those he uses to later describe modernism's characteristic spatiality.[47] Jameson suggests that "space" is a formal symptom of modernist texts in general, because they, too, encounter the representational crisis exemplified in Lewis's *Tarr*: the need to make sense of life in a "metropolis" whose immanent logic—that of imperialism—lies beyond its national borders. As in his discussion of *Tarr*, Jameson's emphasis is on form, even though in his discussion of Forster's *Howards End*, the term "national allegory" is not used. It is significant that in the reemergence in the third world of what was described as an "outmoded" category by the time of the Soviet Revolution, Jameson's discussion of national allegory is no longer posed in terms of the work of form on specific "aesthetic dilemmas," nor in the form of a "representational crisis" that involves and invokes the bounded space of the nation. Instead, national allegory names the condition of possibility of narration itself in the third world. It names it, further, as a *positive* condition, one in which there remains a link, however threatened, tenuous, and political, between the production of narrative and the political. This connection in the first world has been shattered so completely that third-world texts appear "alien to us at first approach."[48]

What I think this suggests is that the nation has disappeared from third-world national allegories. What Jameson describes as national allegory could just as easily have been called political allegory: the nation seems to serve little purpose here, and can only inhibit analyses of third-world literary texts insofar as it seems to point to the nation as the (natural) space of the political in the third world. So again, why *national* allegory? It does not have to do with the historical reemergence of the international system of nation-states—or of the emergence of a new form of this system, which we might too hastily identify as globalization—that formed the "structural and semantic givens" for Lewis at the beginning of this century. Nor does it seem to me that third-world literary texts face the representational problems of modernism: in the third world, lived reality is *never* seen as intelligible only in terms of the "national" situation, and so there is correspondingly no aesthetic or formal necessity to grapple with what amounts to the "absent cause" of lived experience. The "nation" means something else entirely, something different from simply the empirical community or collectivity for which the cultural revolution is undertaken. Jameson's evocation of the nation in his discussion of third-world literature should be taken instead as a reference to a reified "cultural pattern" that "having once been part of the solution to a dilemma, then become[s] part of the new problem."[49]

This is no doubt why the nation has become ever more prominent in Jameson's recent explorations of globalization. For even though it might now seem as if postcolonial literature circulates within a very different set of sociohistorical coordinates than the one that Jameson outlines in "Third-World Literature," the nation remains an ineliminable structural presence within the contemporary "cultural pattern." Far from rendering national allegory useless, globalization makes it an increasingly important interpretive mode or problematic. But here, too, problems arise unless we understand precisely what Jameson means by the nation and how in turn he imagines its relationship to globalization.

The nation has been one of the main sites of struggle in the attempt to understand and conceptualize globalization—whether globalization is understood as the name for a set of real, empirical processes that characterize variously the cultural, social, and economic dimensions of contemporary capitalism, or as the name for a number of competing narratives about the evolving shape of the contemporary political landscape and of the character of any future polity.[50] It has been frequently suggested that globalization has rendered the nation-state irrelevant, because (for instance) the nation no longer seems to retain any juridical power or control over capital or labor, both of which cross borders and evade state surveillance with increasing ease (though far more so in the case of capital and its associated modes of credit, finance, and the like, than in the case of the physical bodies of individual laborers). Then there is the (more or less) antithetical position, which holds that the decline of the nation and nation-state has been much exaggerated. Not only are most companies "tethered to their home economies and . . . likely to remain so," but it is also only the actions of sovereign nation-states that have produced new forms of sovereignty in the form of international regulatory mechanisms like the GATT (General Agreement on Tariffs and Trades) and NAFTA (North American Free Trade Agreement), and also nation-states that have ensured compliance with the global operations of the market at a national level.[51] More recently, commentators have wanted to suggest that neither of these two poles adequately makes sense of the complex, heterogeneous position of the nation-state within globalization. This is, in part, as Jean and John Comaroff point out, because "there is no such thing, save at very high levels of abstraction, as 'the nation-state'": in many polities, neither the "nation" nor the "state" exists as such, while in other places there exists a deep fissure between state and government that makes it impossible to speak of anything that approaches typical ideas about what a functioning nation-state looks like.[52] Put differently, "the processes by which millennial capitalism is taking shape do not reduce to a simple narrative according to which the nation-state either lives or dies, ebbs or flourishes. Its impact is much more complicated, more polyphonous and dispersed, and most immediately felt in the everyday contexts of work and labor, of domesticity and consumption, of street life and media-gazing."[53]

Whether it has died or still lives, the nation-state has long represented the specifically modernist political project of creating citizen-subjects defined through their attachment to national identities. Connected to this project (which on its own it is easy to be suspicious of) is a whole history of left political engagement that has made effective use (or so the story goes) of this historical compromise between capital and labor to bring about the social gains associated with left activism over the past 150 years or so. Whether or not the powers of the nation-state have declined over the past several decades, the nation as such is thus frequently evoked or imagined as the only possible site of progressive politics (due largely, it seems, to its scale) and as thus something that should be fought for in order to maintain or preserve the political project of the left.[54] This desire for the possibilities (incorrectly) associated with the nation-state cannot help but be confused with more empirical analyses of its function within globalization, which is perhaps why the defense of the nation continues to be associated with a left that in the past sought to distance itself from nationalism.[55] Against this position, Michael Hardt and Antonio Negri have strongly asserted that "it is a grave mistake to harbor any nostalgia for the powers of the nation state or to resurrect any politics that celebrates the nation."[56] For them, the relative decline of the sovereignty of the nation-state is the result of a historical, structural process—the globalization of production and circulation, backed up by those supraterritorial agreements that have incurred the wrath of antiglobalization protestors—and is not "simply the result of an ideological position that might be reversed by an act of political will."[57] They also point out that "even if the nation were still to be an effective weapon, the nation carries with it a whole series of repressive structures and ideologies" of which a properly left politics should be appropriately wary.[58] Too-simple demands concerning the political or conceptual necessity of the nation or of the nation-state need to be treated with proper caution, or need to be seen as a potentially debilitating form of nostalgia for political possibilities that no longer exist.

These comments are perhaps somewhat unnecessary; as Comaroff and Comaroff suggest, the scholarly debate over the fate of the nation-state in globalization "has become something of a cliché."[59] I make them here only because in the absence of such ground-clearing or stage-setting, it is possible to mistake Jameson's recent interest in the nation as little more than nostalgia for a modernist form of politics (a politics that believes in the citizen rather than the consumer) in very much the same way that some critics have taken his interest in the third (or indeed, in the second) world as a search for an Other to a capitalism that "has no social goals."[60] A cursory reading of either of Jameson's most explicit attempts to theorize globalization does little to dispel this impression. In "Notes on Globalization as a Philosophical Issue," he laments the "tendential extinction of new national cultural and artistic production" that is the consequence of the domination of the global cultural industries by the United States and endorses state support of culture in places like France and Canada.[61] He also makes the claim that in the first world the powers of the state "are what must be protected against the right-wing attempts to dissolve it back into private businesses and operations of all kinds," a point he reaffirms in "Globalization and Political Strategy," where he states outright that "the nation-state today remains the only concrete terrain and framework for political struggle," even though the struggle against globalization "cannot be successfully prosecuted to a conclusion in completely national or nationalist terms."[62]

While this might seem to be an affirmation of the kind of view of the nation that Hardt and Negri warn against, in the context of Jameson's supple examination of the contradictions and antinomies of globalization a different reason for foregrounding the nation emerges that is of a piece with its presence in his discussion of third-world literature. In both of his recent articles on globalization, Jameson tries to gauge the significance of the global export of American mass culture (through its intersection with the economic, social, and technological) in order to understand what it might mean to try to oppose or to resist its spread around the world. This is, of course, an expression of the cultural imperialist thesis in a nutshell: an understanding of globalization that, while still predominant in the cultural imaginary of academics and the general public alike, has been criticized as misunderstanding the contemporary operations of culture and power.[63] But while on the surface Jameson seems merely to express a Western academic's worries about the disappearance of traditional ways of life, the reappearance of the nation as a conceptual concern complicates our desire to see globalization as something to be either lamented or celebrated. For instance, what Jameson finds disturbing about the global triumph of American cinema is that it marks

> the death of the political, and an allegory of the end of the possibility of imagining radically different social alternatives to this one we now live under. For political film in the '60s and '70s still affirmed that possibility (as did modernism in general, in a more complex way) by affirming that the discovery or invention of a radically new form was at one with the discovery or invention of radically new social relations and ways of living in the world. It is those possibilities—filmic, formal, political, and social—that have disappeared as some more definitive hegemony of the United States has seemed to emerge.[64]

This demand for the persistence of other modes of national culture has little to do with the nation as such. It isn't the case, for example, that

Jameson lauds French film because it is formally or thematically richer than American film, either due to its relationship to some purer national essence (say, summer misadventures in the provinces as an adolescent, an apparently inescapable theme for French filmmakers) or because it is produced outside of the strict demands of the market (as a result of state subsidies). Rather, in our present political and cultural circumstances, the nation names for Jameson the possibility of new social relations and forms of collectivity not just "other" to neoliberal globalization, but the possibility of imagining these kinds of relations at all. Such forms of collectivity are not to be found in some actual national space: "Today no enclaves—aesthetic or other—are left in which the commodity form does not reign supreme."[65] Rather, the nation is now part of the new problem of contemporary cultural revolution, a part of the problematic of globalization that one cannot avoid even if one shares Hardt and Negri's suspicions about the politics of actually existing nation-states; it once again names a reified "cultural pattern," though with different valences and different connections to other concepts and problems than before.

The nation stands for three things in Jameson's recent reflections on globalization. It identifies, first, the possibility of other modes of social life that are organized in strikingly different ways than the American-led "culture-ideology of consumption." Other "national situations" offer models of different forms of collective and social life—not, it is important to add, in the form of "traditional" or "prelapsarian" modes of social being, but in the form of "rather recent and successful accommodations of the old institutions to modern technology."[66] Second, the nation is the name for a frankly utopic space that designates "whatever programmes and representations express, in however distorted or unconscious a fashion, the demands of a collective life to come, and identify social collectivity as the crucial centre of any truly progressive and innovative political response to globalization."[67] These words at the end of "Globalization and Political Strategy" are actually meant to define the word *utopian* rather than *nation*. The link between the two terms is made possible in a note that appears a few pages earlier, where Jameson claims that "the words 'nationalism' and 'nationalist' have always been ambiguous, misleading, perhaps even dangerous. The positive or 'good' nationalism I have in mind involves what Henri

Lefebvre liked to call 'the great collective project,' and takes the form of the attempt to construct a nation."[68]

Finally, Jameson discusses the nation not in order to settle the case either for or against globalization—rejecting, for instance, the false universality of the "American way of life" in favor of one of so many other (rapidly evaporating) national models, which themselves have never yet yielded positive social alternatives—"but rather to intensify their incompatibility and opposition such that we can live this particular contradiction as our own historic form of Hegel's unhappy consciousness."[69] If "Globalization and Political Strategy" ends with a discussion of utopia, "Notes on Globalization" ends with a discussion of the necessity of the dialectic, and of the Hegelian dialectic in particular. The aim of the dialectic is to understand phenomena in order, finally, to locate the contradictions behind them: in Hegel's *Logic*, the discovery of the Identity of identity and nonidentity that reveals Opposition as Contradiction. But this is not the final moment: "Contradiction then passes over into its Ground, into what I would call the situation itself, the aerial view or the map of the totality in which things happen and History takes place."[70] Such a map of the moment when the nation is thought to have been superseded once and for all can only be produced if the nation, the Ground of an earlier moment, is put into play in the dialectic rather than suspended from the outset.

And here we find that we have looped back around to Jameson's discussion of the ineliminable horizon of those objective "cultural patterns" that third-world writers have to confront just as much as first-world critics. Which is a long way of saying that far from obliterating the Marxian problematic, especially with respect to the contemporary use and abuse of culture, globalization makes it more important than ever.

NOTES

I want to thank Nicholas Brown, Caren Irr, and Susie O'Brien for their helpful comments on an earlier version of this article.

1. Jameson, "Third-World Literature in the Era of Multinational Capitalism," 69.
2. Ahmad, *In Theory*, 95–122.
3. Though there has been a good deal of criticism of Jameson's reading of third-world literature, he has also drawn support for his attempt to offer an abstract, general model of literary production in the colonial and postcolonial world. Jean Franco has

suggested that Jameson's generalizations are useful because they "provoke us to think of exceptions." Franco, "The Nation as Imagined Community," 131. With respect to contemporary cultural production in India, Geeta Kapur writes that "Jameson's formulation about the national allegory being the pre-eminent paradigm for Third World literature continues to be valid . . . the allegorical breaks up the paradigmatic notion of the cause . . . it questions the immanent condition of culture taken as some irrepressible truth offering." Kapur, "Globalisation and Culture," 24–25. Michael Sprinker has misgivings about some of Jameson's claims, but finds nevertheless that he puts forward a "provocative hypothesis" that needs to be carefully considered: "Is it not possible, as Jameson here maintains, that certain forms of collective life have until now persisted more powerfully outside the metropolitan countries? And if this be so, of what value are these, perhaps residual but still vital forms of social practice?" Sprinker, "The National Question," 7–8.

4. It is important to recognize just how foreshortened the history of postcolonial studies is within academic discourse. For instance, two of the formative essays in the field, Spivak's "Can the Subaltern Speak?" and Bhabha's "Signs Taken for Wonders," were published in 1985. Jameson's essay is roughly contemporaneous with these essays and should be taken as an attempt to situate Marxist criticism within the general problematic being developed within postcolonial studies at the time.

5. All uses of the terms first world and third world should be understood, following Santiago Colás's suggestion, as being used *sous rature* so as to mark "both the inadequacy and the indispensability of the terms and the system of geopolitical designations to which they belong." Colás, "The Third World in Jameson's *Postmodernism or the Cultural Logic of Late Capitalism*," 259.

6. Seshadri-Crooks, "At the Margins of Postcolonial Studies: Part I," 19.

7. This is essentially the critique that Spivak makes of Jameson's theory of the postmodern in her *Critique of Postcolonial Reason*, 312–37. See also Dipesh Chakrabarty's challenge to the "politics of historicism" in *Provincializing Europe*.

8. Jameson, *Postmodernism, or, the Cultural Logic of Late Capitalism*, 15.

9. For example, Philip Darby has pointed to the failure of postcolonial theory to engage with international relations theory in *Fiction of Imperialism*. Chakrabarty's analysis of the politics of historicism, including those historicisms such as Ernst Mandel's and Jameson's, which remain indebted to Marx's placement of capitalism at the leading edge of historical time, foregrounds the theoretical problems that arise in attempts to think a global totality. While he is right to criticize the Eurocentrism of historicism, the difficulty of developing a different model of history that doesn't reduce it to "sheer heterogeneity" can be seen in his unproductive attempt to develop an alternative model of historicity that enables one to "think about the past and the future in a nontotalizing manner" only by passing through the ontological dead zone of Heidegger's thought. Chakrabarty, *Provincializing Europe*, 249.

10. Hardt and Negri, "Totality."

11. "Every individual interpretation must include an interpretation of its own existence, must show its own credentials and justify itself: every commentary must be at the same time a metacommentary." Jameson, "Metacommentary," 10.

12. Jameson writes that "this whole talk aims implicitly at suggesting a new conception

of the humanities in American education today." Jameson, "Third-World Literature," 75. In his response to Ahmad's criticisms, Jameson states at the outset that "the essay was intended as an intervention into a 'first-world' literary and critical situation, in which it seemed important to me to stress the loss of certain literary functions and intellectual commitments in the contemporary American scene." Jameson, "A Brief Response," 26.

13. Jameson, "Third-World Literature," 69.

14. Jameson, "Third-World Literature," 69.

15. This is one of Ahmad's major criticisms of Jameson. By utilizing the "Three Worlds Theory" as his primary interpretive matrix, Ahmad suggests Jameson is unable to see that capitalism, socialism, and colonialism are all present within the third world. Colás also points out that there are "not only many 'Third Worlds' and many 'First Worlds'; but there are also 'Third Worlds' within the 'First World' and vice-versa." Colás, "Third World," 259. It is worth mentioning here Colás's examination of the paradoxical function of the third world in Jameson's *Postmodernism, or, the Cultural Logic of Late Capitalism*, which is more or less repeated in his essay on third-world literature: "It is *both* the space whose final elimination by the inexorable logic of late capitalist development consolidates the social moment—late capitalism—whose cultural dominant is postmodernism, and the space that remains somehow untainted by and oppositional to those repressive social processes which have homogenized the real and imaginative terrain of the 'First World' subject." Colás, "Third World," 258.

16. Jameson, "Third-World Literature," 69. This claim, which can be redescribed as the loss of any genuinely historical thinking in the postmodern period, is one of the repeated themes in Jameson's work.

17. Sprinker suggests that "we may wish to inquire, are First World allegorical forms so utterly unconscious of their potential transcoding into political readings? Leaving aside the whole rich territory of contemporary science fiction, about which Jameson himself has taught us so much, what about so-called film noir? Surely Fritz Lang, Billy Wilder, and the other émigrés who pioneered this form understood perfectly well that they were making sociopolitically coded films. On the contemporary scene, there is the massive presence of Francis Ford Coppola, not to mention David Lynch, filmmakers whose affinities with the supposedly disreputable mode of social allegory Jameson has discussed with great insight." Sprinker, "The National Question," 6. It is probably possible to cite endless counterexamples in this way; and yet it is important to note that this is to have somehow missed Jameson's fundamental point entirely.

18. Jameson, "Third-World Literature," 69.

19. Jameson, "Third-World Literature," 73.

20. Jameson, "Third-World Literature," 73.

21. Jameson, "Third-World Literature," 85.

22. Jameson, "Third-World Literature," 71.

23. Jameson, "Third-World Literature," 70.

24. Jameson has noted that "it is very easy to break up such traditional cultural systems, which extend to the way people live in their bodies and use language, as well as the way they treat each other and nature. Once destroyed, those fabrics can never

be recreated. Some third-world nations are still in a situation in which that fabric is preserved." Jameson, "Notes on Globalization as a Philosophic Issue," 63.

25. Johannes Fabian has described this "time lag" as "allochronism"—a denial to the "other" of any possible contemporaneity with the West. See Fabian, *Time and the Other*. See also my discussion of allochronism in the Canadian context: Szeman, "Belated or Isochronic?: Canadian Writing, Time and Globalization."

26. Dipesh Chakrabarty's work has engaged directly with the need to simultaneously "think" and "unthink" modernity in the conceptualization of third-world histories and third-world politics; see his *Provincializing Europe*, especially chapter 1.

27. Jameson, "Third-World Literature," 72.

28. See Jameson, "Modernism and Imperialism."

29. Jameson, "Third-World Literature," 72.

30. Jameson, "Third-World Literature," 76.

31. Jameson, "Third-World Literature," 76.

32. Jameson, "Third-World Literature," 78.

33. Jameson, "Third-World Literature," 78.

34. Adorno, "Theses on the Sociology of Art," 121.

35. Julie McGonegal has shown how Jameson's mode of national-allegorical interpretation reveals narratives that reading strategies focused on "Manichean" allegories cannot. Part of her point is that critics of Jameson have confused his elaboration of an interpretative hermeneutic ("third world texts are . . . *to be read* as national allegories") with the thing itself (third-world texts are national allegories, the nation still has significance in the third world, the third world is homogeneous, etc.), and in so doing have missed his metacritical emphasis on the way in which third-world texts necessarily appear to first-world readers as "already read." See Julie McGonegal, "Post-Colonial Contradictions in Tsitsi Dangaremba's *Nervous Condition*: Toward a Reconsideration of Jameson's National Allegory," unpublished manuscript.

36. Raymond Williams provides an account of the historical development of the concept of literature in *Marxism and Literature*.

37. I elaborate on this rather abstract formulation in my book *Zones of Instability*.

38. This is intimated in the final footnote of "Third-World Literature": "What is here called 'national allegory' is clearly a form of just such a mapping of the totality, so that the present essay—which sketches a theory of the cognitive aesthetics of third-world literature—forms a pendant to the essay on postmodernism which describes the logic of cultural imperialism of the first world and above all of the United States." Jameson, "Third-World Literature," 88 n. 25.

39. Jameson, *Fables of Aggression*, 90.

40. Jameson, *Fables of Aggression*, 90–91.

41. Jameson, *Fables of Aggression*, 93.

42. Jameson, *Fables of Aggression*, 94.

43. Jameson, *Fables of Aggression*, 94.

44. Jameson, *Fables of Aggression*, 94.

45. Jameson, *Fables of Aggression*, 94.

46. Jameson, *Fables of Aggression*, 95.

47. See Jameson, "Modernism and Imperialism."

48. Jameson, "Third-World Literature," 69.
49. Jameson, "Third-World Literature," 78.
50. For a perceptive taxonomy of the latter, see Hardt, "Globalization and Democracy."
51. Hirst and Thompson, *Globalization in Question*, 2.
52. Comaroff and Comaroff, "Millennial Capitalism," 325.
53. Comaroff and Comaroff, "Millennial Capitalism," 325. The complexities that exist here can be seen in the way in which globalization itself sometimes provides the basis for the reconstitution or concentration of national energies. Frederick Buell has suggested recently that in the United States globalization seems to be a form of "cultural nationalism for postnational circumstances." Buell, "Nationalist Postnationalism," 550. R. Radhakrishnan makes a similar point when he suggests that "postnational developments are never at the expense of nationalist securities; if anything, they foundationalize nation-based verities and privileges to the point of invisibility." Radhakrishnan, "Postmodernism and the Rest of the World," 42.
54. See Brennan, "Cosmo-Theory."
55. See Rosa Luxemburg, *The National Question*. In Canada, for example, left nationalism represented by groups such as the Council of Canadians seems to have experienced a revival within the antiglobalization protest movement more generally.
56. Hardt and Negri, *Empire*, 336.
57. Hardt and Negri, *Empire*, 336.
58. Hardt and Negri, *Empire*, 336.
59. Comaroff and Comaroff, "Millennial Capitalism," 318. The same could be said for other attempts to fix the particular spaces in which globalization is played out, although the very best discussions of the function of regionalism in globalization or of the new role played by cities does contribute to our understanding of the "polyphonous and dispersed" impact of globalization. See for example Ching, "Globalizing the Regional, Regionalizing the Global"; Mbembe, "At the Edge of the World"; and Sassen, "Spatialities and Temporalities of the Global," among others.
60. Jameson, "Globalization and Political Strategy," 62.
61. Jameson, "Notes on Globalization," 61.
62. Jameson, "Notes on Globalization," 72; Jameson, "Globalization and Political Strategy," 65–66.
63. See especially Tomlinson, *Cultural Imperialism*. For an ethnographic consideration of the limits of the cultural imperialist thesis, see Watson, ed., *Golden Arches East*.
64. Jameson, "Notes on Globalization," 62.
65. Jameson, "Notes on Globalization," 70.
66. Jameson, "Notes on Globalization," 63.
67. Jameson, "Globalization and Political Strategy," 68.
68. Jameson, "Globalization and Political Strategy," 64 n. 11.
69. Jameson, "Notes on Globalization," 64.
70. Jameson, "Notes on Globalization," 76.

CHAPTER 2

Culture and Globalization, or, The Humanities in Ruins (2003)

It is self-evident that nothing concerning art is self-evident anymore, not its inner life, not its relation to the world, not even its right to exist. The forfeiture of what could be done spontaneously or unproblematically has not been compensated for by the open infinitude of new possibilities that reflection confronts. In many regards, expansion appears as contraction. The sea of the formerly inconceivable, on which around 1910 revolutionary art movements set out, did not bestow the promised happiness of adventure. Instead, the process that was unleashed consumed the categories in the name of that for which it was undertaken.

—Theodor Adorno, *Aesthetic Theory*

The nation understands itself as its own theme park, and that resolves the question of what it means to live in Italy: it is to have been Italian once.

—Bill Readings, *The University in Ruins*

As the range and number of books and articles exploring culture in the era of globalization should indicate, the concept of culture has undergone a significant change at the end of the twentieth and in the early twenty-first centuries—a shift that has necessitated new ways of thinking and writing about culture.[1] This is not only, or even primarily, due to the impact on culture of those forces now inextricably associated with globalization: the unprecedented intensification and extensification of electronically-mediated culture on a world-wide scale; the effects of the growth of finance capitalism, that is, of obsessive speculation on capital itself in place of the attention once paid to the products of industry; a political shift from nation-state based sovereignty to a diffusion of sovereignty into international organizations, trade conventions, nongovernmental organizations (NGOs), and transnational corporations; and so on. While these forces, individually and collectively, *have* changed culture and cultures, what is more significant is the conceptual impact of these (thus far largely) empirical developments. Early work on globalization tended to claim that it constituted something like a genuine historical and epistemic break: on the other side of 1989 (the beginning of the end of the Soviet Empire), everything is supposedly different. It has now become more common to see through the rhetoric of newness that surrounds globalization, and to insist on the development of these forces in the *longue durée*. As with the economy and politics, so too with culture: rather than creating anything genuinely "new" in the sphere of culture, globalization has produced the conditions that might permit us to rethink culture in a larger historical frame, a process that would allow us to see that the concept of culture has *always been other* than what it claimed to be.

But if globalization has raised this possibility, its actualization has been repeatedly blocked by the operations of culture itself. The typical discussions that emerge around culture in reference to globalization—the already tired talk of cultural mixing-and-matching, or the equally unoriginal worry about the threats (and possibilities) posed to this or that culture by (American) mass culture—merely continue the old game of culture in a new guise. What is original about globalization for culture is *not*, it seems to me, to be found in the sudden impact of cultures upon one another. Rather, it is that globalization has made it impossible to maintain any of the fictions that have continued to circulate around the Western concept of culture. This can be seen most acutely, I think, in the current crisis facing the humanities, which is why any exploration of culture and globalization must ask the question of what globalization means for the humanities today and for the future. But before we can address this question, we need to consider the ways in which the concept of culture has typically circulated in and alongside globalization discourses, in order to understand what is missing in most explorations of culture in the era of globalization.

Culture and Space

Discussions of globalization and culture have typically focused on the way in which both physical and immaterial *speed*—the movements of goods and people, as well as money and electronic signals—has reconfigured the *space* of culture. In the study of national literatures or histories, languages or cultural traditions, or any form of what used to be referred to as "area studies," culture has long been intimately related to geography. Even though it has also always been clear that culture must be understood as fluid and unbounded, as something able to travel and exert its force across boundaries, culture has nevertheless been understood primarily as something that exists in fixed, determinate spaces, whether this is the space of the nation and the region, villages, groups, or subcultures. Since at least the nineteenth century, and in conjunction with the solidification of the nation as a political form, there have been repeated attempts to define and differentiate national culture and character (from Johann Gottfried Herder to Hippolyte Taine, from Fred Morley to Fred Lewis Pattee).[2] Though the shaky logic of national culture has been repeatedly challenged, these theoretical linkages between culture and geography have persisted as a powerful conceptual commonplace, appearing as the subject of an annual deluge of nonfiction books investigating the national character (for example) of the United States and Canada, as well as forming the basis of countless travel narratives and the animating substance of journalistic reportage. In the wake of 9/11, what Theodor Adorno referred to as "the detestable jargon of war that speaks of the Russian, the American, surely also of the German"[3] has experienced a notable resurgence in the form of a populist, Orientalist discourse of the "clash of civilizations" between the West and Islam, which has further reinforced the idea of absolute cultural divides between peoples, based on what Taine referred to as "race, moment and milieu."[4]

Even though these recent anti-Islamicist discourses suggest that less has changed than one may have thought, the speed associated with globalization has been connected (for better and worse) to the obliteration of the spaces in which culture was once thought to "naturally" or "normally" dwell, as well as to the destruction of the borders that were once imagined as marking cultures off from each other. In the era of globalization, cultural boundaries are imagined as having become porous, indefinite, and indeterminate: the "local" intersects with the global (and vice versa), and culture becomes unsettled, uprooted, hybrid, mixed, and impure. Globalization is the moment of mass migration, multiculturalism, and cosmopolitanism; if the nation was once imagined as a community through the aid of newspapers and novels, the ubiquity of new forms of mass culture has led to new, transnational regimes of the imagination. With respect to culture, discourses of globalization are thus often focused on border zones, and on the complex negotiations that take place as these borders are explored, reimagined, and reasserted in a world of increasing, if unequal, cultural interaction. Much of the analysis of borders has focused rightly on the implications of these power differentials (differentials of scale as well as speed) on the form that these cultural interactions take. As problematic as the discourse of cultural imperialism has been, discussions of the globalization of culture in both academic and public spheres nevertheless continue to imagine the conjunction of these terms as a narrative about "Americanization," or of the threat posed by Western cultural products to the cultural autonomy of non-Western, still-modernizing communities and regions.[5] A direct line can be drawn from one of the first major works on cultural imperialism, Ariel Dorfman and Armand Mattelart's *How to Read Donald Duck* (1975), to the authors of the recently published *Key Concepts in Post-Colonial Studies*, who claim that "the key to the link between classical imperialism and contemporary globalization in the twentieth century has been the role of the United States," which is responsible for initiating "those features of social life and social relations that today may be considered to characterize the global: mass production, mass communication and mass consumption."[6]

What is interesting is that while there have been repeated claims that globalization produces new conditions for culture—new and unprecedented forms of cultural intermingling and interconnection that, in Canada at least, is celebrated as the coming-into-being of a paradoxically ethnicized post-ethnic state—culture is *still* imagined in virtually all of these formulations as connected to geography in a more or less Romantic fashion. After all, globalization can only pose a threat to cultural autonomy if cultures are conceptualized as being necessarily (for purposes of individual and collective self-identity) autonomous in the first place. The reason it is possible for discourses of cultural mixing (as in multiculturalism) and radical cultural otherness (as in the sweeping and uncritical return of Eurocentrism in the current war on terrorism) to exist side-by-side in globalization is that, to a large degree, the former presumes the latter: hybridity necessitates conceiving of cultures as monadic to begin with, whether historically or conceptually, or both. While culture is thought to have entered a new situation in globalization, it seems to me that the concept of culture itself *hasn't* undergone a similar change or shift. The conceptual boundaries within which culture is able to move remain those first delimited by Herder and Taine centuries earlier. Globalization has forced theorists to think seriously about the implications of the dislocation or deterritorialization of culture, and to try to think about culture after its ties to blood, belonging, and soil have been severed. But it seems that most attempts to conceptualize what globalization means for culture have only gone halfway: once disembedded from geography, the function and meaning of culture needs to be redefined in a radical way if the concept is to continue to have any meaning at all. That this hasn't happened has more to do with institutional and disciplinary inertia than with the continued applicability or utility of the Western idea of culture to the conditions of the global present. Or rather, since the shifting meaning of "culture" has charted "within its semantic unfolding humanity's own historic shift from rural to urban existence, pig-farming to Picasso,"[7] the lack of a shift now needs to be probed and assessed to determine what culture still signifies.

Culture and Time

The contradictions that emerge from the persistence of an older concept of culture in the investigation of the conditions of its dissolution can be seen in the conflicting views that have been expressed—often at the same time—about the temporality of globalization. One of the important (and importantly contested) assumptions of the typical narrative of globalization and culture is that globalization constitutes an historical rupture, a break with the past that inaugurates a new era of cultural relations. This rupture is usually *not* imagined as something completely new, that is, as a whole new *episteme* that marks the end of modernity and the birth of something else. Rather, preexisting tendencies and processes (economic, political, social, etc.) are thought to have simultaneously undergone an epochal intensification. Globalization is imagined, in other words, as that moment on a graph of a logarithmic equation where the line suddenly spikes skyward; it is the moment when this spike occurs everywhere at the same time, if with greater or lesser degrees of intensity. For these reasons, globalization has been employed primarily as a periodizing term, the name for a particular moment in history, though it has by extension also been used to describe the set of processes that have produced, or that are contained in, this moment. These narratives of historical rupture have been accompanied by critiques that have taken the form not of an outright rejection of this periodizing hypothesis, but of attempts to downplay both the intensity and extensivity of globalization through references to historical precedents and the *longue durée*. Such deflationary counternarratives have been articulated in the fields of economics, studies of migration and the interaction of social communities, global politics, and even communication technologies.[8] With respect to culture, these critiques point out that culture and cultural forms have long traveled outside of their "natural" boundaries; that is, that the interaction and hybridization of culture associated with globalization is part of a longer process. As Christopher Clausen has put it, the process of breaking down boundaries between cultures "sometimes misidentified with the electronic age—began long before computers were invented, and whether we label it globalization, modernity, assimilation, cultural imperialism, the technological revolution, or the inexorable logic of capitalism, no culture is immune to it."[9]

The debates over the appropriate historical frame of globalization have significance for the concepts and theories that are employed to make sense of the contemporary world. Theories that envision a historical rupture occurring with Bretton Woods, the Vietnam War, or the end of the Soviet Bloc (and there are, of course, other possibilities) trumpet the need for new concepts, and the reconfiguration or reevaluation of older ones. On the other hand, those that place globalization within a longer history tend to see older theories and concepts as still having utility. With respect to discussions of culture and globalization, *both* of these scenarios have been played out, though along different axes of analysis. In the first instance, new models for the *circulation* of culture have been proposed in order to make sense of the apparently discontinuous spread and impact of contemporary culture, the most well-known being the vocabulary of scapes, flows, and cascades developed by Arjun Appadurai in an effort to understand the "complex, overlapping, disjunctive order" of the new global cultural economy.[10] Even in this case, however, what seems to be untouched by any of the transformations produced by globalization is our understanding of the concept of culture itself. For Appadurai, culture now moves differently, and its new mode of circulation produces new kinds of cultural effects (e.g., "localized" outbreaks of ethnic violence whose root cause lies in the financial support funneled to extremists by extra-local or extra-national migrant communities). Yet even here, culture continues to play the role that it has long performed, acting as the primary site where individual and collective identities are shaped and formed; if anything, his insistence on the new role played by the "imagination" in the global order reinforces a Romantic view of culture, even if he also argues that it is important to "capture the impact of deterritorialization on the imaginative resources of lived, local experience."[11]

To summarize: on the one hand, globalization names a new condition for culture that is related to the sudden dissolution of culture's boundaries and its increased global motility. And yet, the culture that is suddenly mobile and deterritorialized is still imagined largely in its old guise of human expressivity as something strangely (and yet familiarly) unaffected by the hurly-burly of empirical social transformations, or as its opposite: the debased culture of mass culture, now imagined as disastrously writ large over the face of the entire globe, subsuming everything in its path. Yet neither of these concepts of culture seems to adequately express the conditions under which culture is produced and circulated today (much less how culture functions), what this category means or describes, and how it relates to or mediates social life more generally—or even *if* its role is one of mediation any longer.

The Humanities and the "Cultural Turn"

Perhaps counterintuitively, this development is confirmed by the increasing significance of culture in discussions of globalization, and indeed in the social sciences more generally (as witnessed in the innumerable discussions of the "cultural turn" that has placed culture back on the agenda of the social sciences). While the discourse of globalization began in the early 1990s as people focused primarily on economic and political change, culture has since become more and more important in thinking about the meaning and consequences of globalization. There are countless examples that one could draw upon. In perhaps the final suturing of the torn halves of base and superstructure, Fredric Jameson and Lawrence Grossberg have both described globalization as the moment in which the economic and the cultural fold into one another, becoming both empirically and heuristically inseparable.[12] On the other side of the political spectrum, Samuel Huntington's thesis on the "clash of civilizations" affirms in its own way the centrality of culture to an analysis of the new global situation. And what John Tomlinson has usefully described as the "complex connectivity" of globalization is expressed in and through culture in a way that places the register of culture at the center of discussions of globalization. Tomlinson suggests that the complex connectivity of globalization has confused the division of human life into the familiar categories of the social sciences: the economic, the social, the political, the technological, etc. As the point of articulation of all these categories—the site or spaces of "meaning construction [that] informs individual and collective actions"[13]—culture is now championed as the key register within which globalization is both experienced and understood.

Such an interest in culture might suggest that the way is open for the humanities—the traditional site for the study of culture in the university—to reassert their importance. Yet the very opposite seems to have taken place. This is due in part to changes in both the ideology and social function of the university over the past few decades: a transformation of the university from secular clerisy to corporation that has been traced out by Bill Readings, Masao Miyoshi, Mary Poovey, and others.[14] Over the past several decades, the humanities have endured funding cuts, a

decline in student enrollment and interest, and an increasing function-
alization of the curriculum, along with a gradual transformation of its
labor pool into part-time and contract workers. These attacks on the
humanities are not simply the result of indifferent, philistine politicians
who don't understand the importance of the humanities (though such
readings are hard to resist and not without some degree of validity),
nor the fault of humanities professors, who haven't asserted themselves
enough in the public sphere to bring needed attention to the crucial role
their work plays in social life. As surprising as this statement might seem
to those engaged in cultural work today, the nation-state *isn't* opposed to
culture. All one needs to do is to look at recent policy documents to see
that it talks about culture incessantly, and does so in the most Romantic
terms possible. To take just one example, the executive summary of the
February 1999 report of the (Canadian) Cultural Industries Sectoral
Advisory Group on International Trade begins: "Culture is the heart of
a nation. As countries become more economically integrated, nations
need strong domestic cultures and cultural expression to maintain their
sovereignty and sense of identity. . . . Cultural industries shape our society, develop our understanding of one another and give us a sense of
pride in who we are as a nation."[15] While this might sound like discourse
that could have emerged from an old-school humanities department,
the reality is that the model or vision of culture produced in and by the
humanities bears little relationship to the one championed by those globalization theorists for whom culture has become everything, or by the
state for whom "culture is a nation's heart," or indeed by multinational
media conglomerates beset by the crisis of a lack of cultural "content" to
circulate through the communication networks that encircle the globe.
The humanities have become marginalized as a result of their inability
to continue to grasp the concept that they have committed themselves
to understanding: the concept of culture *has* shifted, even if this has
yet to be properly registered by the humanities, or by intellectuals more
generally.

How can this be? Over the past forty years, the legitimacy of the
concept of culture that continues to underwrite the humanities has
been under concerted attack—and not from without, but from *within*
the humanities themselves. Postcolonial studies has drawn attention

not only to the blind spots of the Western academy in considering the culture and cultural production of other peoples, but also to the fundamental role played by culture in imperialism and colonialism. In the Western academy, the development of cultural studies has drawn attention to other blind spots, not the least of which has been the way in which "culture" has been used to exercise and legitimate political domination. For example, in his discussion of the historical context of Matthew Arnold's seminal articulation of the relationship of culture to society, Raymond Williams makes clear the links between the assertion of "excellence and humane values" and Arnold's opposition to the "anarchy" of public demonstrations and protests over the extension of the franchise in Britain. In a similar way, Pierre Bourdieu and Terry Eagleton have exposed the ruse of the aesthetic, showing how aesthetic value names a relation of power rather than the special properties of specific objects (like literary texts or artworks) or dispositions of the subject.[16] For both writers, the university was the site at which one learned appropriate modes of aesthetic distinction and cultural interpretation. This is one of the reasons why, as Étienne Balibar and Pierre Macherey argue, the very concept of "literature is inseparable from an academic or schooling practice that defines the conditions for both the consumption and production of literature."[17]

Perhaps most importantly, in their crucial analysis of the coincident development of both culture and the state beginning in the late eighteenth century as "sites of reconciliation for a civic and political society that is seen to be riven by conflict and contradiction,"[18] David Lloyd and Paul Thomas point to the ideological role that culture was to play in the West: "Culture . . . is not confined in its objects to the artistic, or, more narrowly, the literary, but aims rather at the harmonious cultivation of all the capacities of the human subject at a time when it was already apparent that the division of intellectual and manual labor was increasingly formative of specialized or partial individuals."[19]

This conjunction of state to citizen through the medium of culture was the product of a specific moment in history, a moment that we are now past. In the waning of the importance of the nation-state in the operations of global capitalism (and it has waned, even if the state played a crucial role in instigating and instituting the anti-statist

regime of globalization), there becomes less of a need for a social institution geared toward the production of a national narrative, or of a
discourse that mediates the relationship between the populace and the
state. It is this decline of the university, and of the humanities in particular, that Bill Readings outlines in *The University in Ruins*. He writes
that "since the nation-state is no longer the primary instance of the
reproduction of global capitals, 'culture'—as the symbolic and political
counterpart to the project of integration pursued by the nation-
state—has lost its purchase. The nation-state and the modern notion
of culture arose together, and they are, I argue, ceasing to be essential
to an increasingly transnational global economy."[20] Even as the ideology of the humanities gets spread over an increasingly larger sphere
of concern (as suggested, for example, in Appadurai's appeal to the
imagination), the function that this ideology was supposed to serve
has disappeared, along with the institutions that produced it. It's no
wonder that the concept of culture is now open to all kinds of other
uses, but also that there is so much confusion over its uses, as older
definitions and sensibilities collide with new realities that they are
unable to make sense of by means of it.

The Humanities in Ruins

Potentially, the crumbling of the socio-historical conditions that have
produced the need for this particular ideology of culture—an ideology
that has long masked the operations of social power in metanarratives
of progress, humanity, and Enlightenment—opens the way for a new,
less mystified understanding of culture. At the very least, it opens the
way up for methodologies that have always stressed the need to see
cultural objects in networks or systems of power to assume a more
prominent place in the humanities and in definitions of its role and
function. One way of positioning this shift is to suggest that the *analysis* of culture—that is, of what occurs in the name of culture, of what
forms of power and knowledge pass through those objects, practices,
and experiences that we describe as "cultural"—might replace (*e*)*valu-
ation* as the dominant way of thinking of culture (though one needs to
be careful about an opposition that might suggest that it is possible to

drain "value" or politics out of cultural interpretation in a meaningful, non-ideological way; this is not the intent of this distinction here). The specter of value that has long provided the ground of humanities scholarship could give way finally to the examination of the modes and forms of the productivity of culture; globalization might be what brings culture back down to earth from the heavens, insisting on the immanence of what has long imagined itself as transcendent. From this perspective, what might be most significant about globalization—as concept and as empirical reality—is less the rapidity of the circulation of culture within it, or the intensified intersection of cultures with one another, than the fact that this circulation (and the historical circumstances that enable it) makes it difficult, if not impossible, to imagine any longer that the function of culture and of the humanities is to express and defend the "best that has been thought and known." For what the emphasis on the mobility of culture insists upon is not just that this is a new condition of culture, but that culture has always been uprooted and hybrid. That is, *culture has never been what we believed it to be*; it has always had a different function than the guardians of the humanities would have liked to assign to it.

What has mitigated this radical rethinking of the concept of culture, and thus of a new role for the humanities even in the face of radical critiques of its ideological uses, is yet another aspect of globalization and its relationship to culture. If it has remained possible for the humanities to continue to imagine their role as being "the harmonious cultivation of all the capacities of the human subject," and of the university to maintain (at least in official pronouncements) its "grand narrative . . . centered on the production of a liberal, reasoning subject,"[21] it is because in the humanities, global culture is widely conceived of as commodity culture, a form of culture conceived as constituting an attack on modern subjectivity itself. Instead of asking deep questions about the politics of the humanities and of the ideology of culture that sustains it, the combination of fears about commodity culture, along with a new fear of its dislocation of anything and everything once outside of it on a global scale, has allowed the humanities to assume a role with which they are eminently comfortable: the defender of truth and beauty against a philistinism or barbarism that, having become global, is now more dangerous than ever. It comes as no surprise that it is precisely at this point that there has been a return of a more or less classical discourse on "beauty," as reflected by books such as Elaine Scarry's *On Beauty and Being Just*, Wendy Steiner's *Venus in Exile*, and James Elkins's *Pictures and Tears: People Who Have Cried in Front of Paintings*.[22] But the return of this discourse, and of other books that attempt to reassert value in the face of commodity culture, must be seen as a further symptom of the ruin of humanities, rather than a valiant reclamation of its fundamental task: to express what is best and greatest about (an always unhistoricized) human Being.

Such recourse to Arnoldian or Romantic notions of culture in response to globalization is not only to be found in, for instance, the defense of literature or the fine arts against the encroachment of a predominantly visual consumer culture. It is possible to find it as well in forms of apparently more political or politicized discourse, in which what is opposed to mass culture are those aspects of the subject and state that only high culture makes possible (or so it's asserted). With respect to the subject, this concerns the possibility of reason or critical thinking, which in turn is related to the possibilities of citizenship and civic virtues—a common enough connection of subject to state, from Kant to Habermas. The humanities thus come to stand as guardians of critique itself, defenders against a barbarism characterized not by industrial culture and profit (as it was for Arnold), but by an interest in mass culture (expressed paradigmatically in the form of that evil called television). For example, Mark Crispin Miller, a former professor of English who has since become one of the most virulent critics of contemporary media, offers the following account of the decline of critical thinking:

By the mid-Seventies, however, there was one demographic group now "totally into it" [television] for the first time: America's undergraduates, who watched much more and knew much less than any of the student cohorts that had preceded them. So it seemed, at least, to those of us now teaching. No longer, certainly, could you assume that your lit classes would recognize, say, Donne's *Holy Sonnet XIII*, or the *Houyhnhnms*, or the first sentence of *Pride and Prejudice,* or any of the other fragments that had once been common knowledge among English majors.[23]

For Miller, the problem has as much to do with the decline of *reading* as with the lack of knowledge of English literary history:

> Spectatorial "experience" is passive, mesmeric, undiscriminating, and therefore not conducive to the refinement of the critical faculties: logic and imagination, linguistic precision, historical awareness, and a capacity for long, intense absorption. These—and not the abilities to compute, apply or memorize—are the true desiderata of any higher education, and it is critical thinking that can best realize them.[24]

Such arguments are common enough. What is more interesting than whether they have any critical bite or not is the way in which a certain vision of the critical faculties, itself a product of history rather than nature, is reified as the one and only mode of real critique. With images of the classical moment of the bourgeois public sphere dancing in their heads, the present can't help but seem like a wasteland to critics who measure the twenty-first century by a whitewashed version of the nineteenth. What such critiques fail to do, of course, is to offer an account of just what function culture performs *now.* Instead, they oppose contemporary culture with their own (already problematic) vision of culture, which they take as truth in much the same way as, in a different context, György Lukács insisted on the political virtues of the realist novel in comparison to its decadent modernist counterparts. Bertolt Brecht's response to Lukács is appropriate in this case, too: it's not the good old days that we should be fascinated with, but rather the bad new ones, and in these bad new days, new forms of culture must necessarily replace the old ones.

Humanities Without Value, Culture Beyond Culture

The bad new days need not be so bad as they are usually thought to be (or maybe the right way to say this is that the present is always bad for those who have to live it). The typical link between globalization and culture tends to obscure, first, the degree to which globalization has disturbed the concept of culture, and second, its impact on the humanities. Globalization has left the humanities in ruins, conceptually and materially. But there are two ways to think about these ruins. One is to see them as a sign of the lamentable end of forces and modes of being essential to democratic life and genuine individual experience; another is to see globalization as opening up the possibility for thinking about contemporary experience and culture in a more complex way than this defensive reassertion of the modern subject and state suggests—that is, as paving the way for a new form of critical humanities that is able to think about culture from perspectives adequate to the age. I have tried to argue thus far (however sketchily) for the limits of the former and the necessity of the latter.

What form would this new critical humanities take? And what role would the concept of culture play within it? Can the humanities do without the array of concepts that it has long associated with culture—concepts such as "genius," "imagination," "creativity," "beauty," and "value"? In what way would such a practice continue to be the humanities? (And why is it still necessary to address these same old questions?) It is admittedly more difficult to answer these questions than it is to identify the problematic circulation of an older vision of the humanities in the new circumstances of globalization. But at least the outlines of a critical humanities are, I think, easily grasped. There are no absolute beginnings: a humanities that takes seriously the analysis of its historical and ideological genesis will still have to draw on this history to make sense of the ways in which historical developments have reconstituted the grounds of its own practice. The humanities would continue to be defined as a practice that explores culture, but one that takes as a central principle of its practice the notion that culture is constituted in entirely different ways at specific moments in time. Strangely, contemporary literary history (for instance) has been better at achieving this than have studies of the contemporary moment itself. There are clear models for such a practice, including Walter Benjamin's "The Work of Art in the Age of Mechanical Reproduction," which begins with a challenge to the categories of "genius, eternal value and mystery" in the arts, and

Pierre Bourdieu's analysis in *The Rules of Art* of the emergence of the cultural sphere in its modern sense in nineteenth-century France.[25] By destabilizing the grounds of the humanities, globalization opens up the possibility of *generalizing* these kinds of critical practices, of moving them from the periphery to the core of the humanities' self-identity.

There is a great deal more that could be said here, but let me end by pointing to some of the theoretical grounds for this new humanities. In order to take advantage of the opening that globalization provides for a new conception of culture, I would like to highlight four interrelated dimensions along which the humanities have to reconsider their theoretical orientations and interpretive practices.

First, those involved in the study of culture need to think seriously about the problem of "affirmative culture," which arises out of the tendency to focus on objects (specific literary and cultural texts, cultural producers, genres, etc.) rather than cultural processes. Affirmative culture is a concept developed by Herbert Marcuse, who described it as the product of a process

> in which the spiritual world is lifted out of its social context, making culture a (false) collective noun and attributing (false) universality to it. This . . . concept of culture, clearly seen in expressions such as "national culture," "Germanic culture" or "Roman culture," plays off the spiritual world against the material world by holding up culture as the realm of authentic values and self-contained ends in opposition to the world of social utility and means. Through the use of this concept, culture is distinguished from civilization and sociologically and valuationally removed from the social process.[26]

It is not only traditional forms of humanistic study that affirm culture in this way: cultural studies, too, has a tendency to oppose culture to the world of utility in the same manner. This is why the distinction between analysis and evaluation that I made earlier can't be taken as a solution to our current impasse, but should be seen as an identification of two positions that in the end are equally unsatisfactory. While proclaiming to study the "everyday," the life of the popular and the mass, cultural studies nevertheless imbues the cultural commodities that it studies with a more traditional "cultural" character through its very insistence on the authenticity of nontraditional cultural forms. As Readings perceptively points out, "cultural studies does not propose culture as a regulatory ideal for research and teaching, so much as seek to preserve the structure of an argument from redemption through culture, while recognizing the inability of culture to function any longer as such an idea."[27] Furthermore, by accepting commodity culture as culture, and by consequently affirming the spiritual dimension of this culture as a site of meaning and significance, cultural studies circulates in a perpetual present in which the reality of present-day culture amounts to no more and no less than all that culture is and can be. The cultural past, dominated by what cultural studies considers to be the lumbering dinosaurs of bourgeois high culture, is closed off from it—but so is the future, since the present of culture is taken as fate. A critical humanities will have to sidestep both traditional humanities study and cultural studies by focusing not on authenticity, but on the social process in and through which cultural objects are produced, circulated, and consumed.

Second, a critical humanities that wants to understand the contemporary function of culture needs to take commodities and consumerism seriously—not as deviations of some true idea of culture, and not primarily as a normative issue (shopping as bad, destructive, etc.), but as a significant transformation in the concept of culture that has had implications that we don't yet completely understand. It has become a critical commonplace to lament consumerism and commodity culture; indeed, it often seems that much of the energy of the humanities emerges out of this lament and the frequently made opposition between (for instance) reading and watching. But such laments fail to interrogate the conditions of consumer culture, being satisfied instead with the presumption that consumerism is either without culture, or its very opposite.

Taking consumerism seriously doesn't imply the negation of a politics of consumerism or consumption—of the kind outlined by (for example) Juliet Schor, who has explored the consequences of (among other things) the growing "aspirational gap" in US society.[28] It remains important to draw attention to the ways in which contemporary mass culture constitutes a concerted form of "public pedagogy"—a pedagogy of hopes, desires, beliefs, and identities—that now outweighs anything that might be taught in schools or homes. Henry Giroux in particular has articulated this point tirelessly in his work on education and mass culture. However, when these critiques devolve into demands for the reassertion of the now lost public sphere, or place hope in the reformation of collectivities of an older kind, the contemporary terrain of culture is dangerously misread. Analyses of consumerism almost always get confused with the normative claims they also want to advance: a clear understanding of how consumer culture operates, for instance, is almost always blurred by the wish that things could be different than they are. It has become nigh impossible to suggest, for instance, that consumerism is itself political—not, in other words, the "other" of civic possibilities and virtues, but an example of their mutation into a radically different form. For all its problems, Néstor Garcia Canclini's claim in *Consumers and Citizens* that "consumption is good for thinking" has the effect of shaking up our pre-established sense of what consumerism is about. "To consume," he writes, "is to participate in an arena of competing claims for what society produces and the ways of using it."[29] Instead of imagining consumers and citizens as existing in an inverse relationship to one another, Garcia Canclini suggests that we investigate consumption as a site "where a good part of economic, sociopolitical, and psychological rationality is organized in all societies."[30] Whether or not Garcia Canclini is right in his sense of how consumption operates, an understanding of culture in the era of globalization cannot avoid seeing consumption as a site of rationality and of cultural experience that, whatever one thinks of it, has a constitutive role to play in contemporary culture.

Third, even after all of the explorations of the ideologies of the humanities, there remains a need for a more thorough investigation of the historical narratives that have legitimized the standard view of culture in the humanities. This is especially true of the narratives that established the modern sense of the mission of the humanities. One such narrative concerns the opposition of modernism to mass culture, an opposition that has elevated the monuments of modernism into exemplary expressions of a critique of the existing world within the realm of art and literature. The narrative that links modernism to revolution has transformed much of the writing on modernism into an elegy over lost political possibilities. This narrative has been challenged recently in Miriam Hansen's writings on "popular-reflexivity" of early cinema, and Susan Buck-Morss's explorations of the unexpected links between Soviet and American twentieth-century popular culture.[31] Perhaps most forcefully, Malcolm Bull has argued that while modernism may have been against *modernity*, it was never against *capitalism*, which is evidenced in part by the seamless assimilation of modernist culture into museums and literary canons.

Bull claims that "modernists were not partisans resisting the present and pressing on eternity, they were negotiating the equally tricky but rather more mundane path between the two cultures of capitalism"—classicism and commodity culture. Rather, "working between two antithetical cultures meant that resistance to the one almost always involved some degree of complicity with the other."[32] But his argument goes beyond the not-uncommon assertion of modernism's incomplete rejection of either classicism or commodity culture. Bull suggests that modernism has to be seen as belated, as working a divide between one culture of capitalism and another that by the beginning of the twentieth century had already been crossed over once and for all:

> For most people, the culture of modernity has been the culture of commodities; or, to put it more bluntly, "postmodernism" was the culture of modernity all along. This is true not just for the huge numbers of people in the twentieth century whose first experience of anything other than folk traditions has been American-style TV; but also for their predecessors who moved straight from agrarian communities to the world of the newspaper and the wireless. . . . Only for those steeped in the classical tradition did postmodernism require new forms of attention.[33]

Such re-narrativizations can help to dissipate what have become unproductive conjunctions between art and culture. It's not that Bull's arguments eliminate the political productivity or engagement of certain forms of modernist cultural production. Rather, by showing us a modernism that is always already contaminated by its historical situation, he helps us to avoid lamenting the irretrievable loss of this moment of supposed purity, which in turn prompts us to look at the politics of culture in our moment as one that not only needs not, but cannot be free of ideological contagion.

Finally, humanities scholars need to reconsider the history of recent theory as reactions to historically specific circumstances that may no longer hold today. When Hardt and Negri describe both post-colonial and postmodern theory as *symptoms* of the end of modern sovereignty—as kinds of critique that can only emerge once modern sovereignty is no longer the framework for control and domination—they do so not in order to deny the utility and importance of many of their formulations.[34] They are pointing, rather, to the way in which any theory expresses incompletely the moment that it is trying to analyze, relying on concepts and narratives that no longer, or incompletely, relate to empirical circumstances. The progressivist narrative in which we have tended to view theory, in which one theory builds on another and we slowly get closer and closer to the truth, tends to obfuscate the historicity of theory itself. Of course, the historicity of concepts is a central element of contemporary theory, such that no one who engages in theory would understand what they do as a project involving truth. Still, in the actual practice of theory, this fact is more often than not lost, and theory becomes yet another narrative of modernity (which means, for instance, that there are more- and less-developed theoretical regions in the world, that theory can be imported from one country to another in the manner of high technology, or even that there can be strategies of import-substitution in the theoretical field).

In making these statements about what the humanities need to do to reinvent themselves in the context of globalization, I don't mean to advocate any particular methodology or interpretive practice. I merely want to suggest this: if the role of the humanities is to explore and to understand the circulation of forms of symbolic and cultural production, if its task is to bring to the surface of social consciousness normally latent processes that take place in these forms—and to do so in a critical fashion, rejecting the commonplaces of the day—they need to direct themselves to the ways in which the profound transformation in the circulation of culture that we have called globalization has also been accompanied by a profound transformation in culture itself. While there has been a great deal of attention paid to the new conditions for the circulation of culture, there is little movement to reimagine the concept of culture as such. This is not a demand for that most precious of commodities—a whole new theory of culture—but a suggestion that one way forward is to reassert or reaffirm those theories that have long drawn attention to the shape of our ideologies of culture, while also giving up on the identity of the humanities as the guardian of the good against commodity culture and commodity aesthetics.

And this is more difficult than it might seem. Pierre Bourdieu made it part of his life's work to deny the importance of the aesthetic, focusing, for example in *The Rules of Art*, on a "scientific analysis of the social conditions of the production and reception of a work of art" (xix) while never once addressing the question of value. However, in the attack that he launched on neoliberalism over the last part of his life, an attack based on the pernicious influence of the logic of neoliberalism over all social spheres, Bourdieu reverted to a vocabulary in which he defended (for example) the production of the great works of European literature, claiming that such masterpieces could only continue to be produced if the fields of cultural production were allowed to remain semi-autonomous.[35] The spread of the logic of neoliberalism across society (measurable, for instance, in the widespread application of the vocabulary of market efficiency in the operation of non-market sectors) demands a response. But is an appeal to aesthetic value an appropriate one? Such an appeal is at best a contradictory one, and one that cannot be seen to really oppose the cultural conservatism that makes up (in its own contradictory way) the dynamism of neoliberalism. What would be better would be a challenge that did not make recourse to the aesthetic at all, but that made an argument within the logic of contemporary culture; but since such a logic has yet to be mapped out, it is not surprising that the critics like Bourdieu remain stuck with a concept of culture that is no longer our own.

NOTES

I want to thank Richard Cavell for giving me the chance to present an earlier version of this chapter at the International Canadian Studies Centre at the University of British Columbia, and Nicholas Brown for offering pointed criticisms and observations about its limits. This chapter was written with the assistance of a Social Sciences and Humanities Research Council of Canada Standard Research Grant.

1. See Chambers, *Culture after Humanism* and Tomlinson, *Globalization and Culture* as just two examples of a large genre of books and articles in this field.

2. Morley and Pattee offered early and influential definitions of the fields of English and American literature. Morley described the connection between English literature and the English nation in 1873 in the following terms: "The literature of this country has for its most distinctive mark the religious sense of duty. It represents a people striving through successive generations to find out the right and do it, to root out the wrong and labor ever onward for the love of God. If this be really the strong spirit of her people, to show that it is so is to tell how England won, and how alone she can expect to keep, her foremost place among nations." The first professor of American literature, Fred Lewis Pattee, began his introductory text on the subject with a description of the relationship between literature and the nation that by the end of the century had become all but indisputable: "The literature of a nation is the entire body of literary productions that has emanated from the people of the nation during its history, preserved by the arts of writing and printing. It is the embodiment of the best thoughts and fancies of a people." Morley and Pattee, cited in Clausen, "National Literatures," 64 and 65 respectively.

3. Adorno, *Aesthetic Theory*, 205.

4. Taine, *History of English Literature*. Adorno suggests that such nationalist thinking "obeys a reifying consciousness that is no longer really capable of experience. It confines itself within precisely those stereotypes that thinking should dissolve." Adorno, "On the Question: 'What is a German?'", 205. This is an apt description of the work of many pundits on 9/11, including Samuel Huntington, writer Robert Kaplan, *New York Times* columnist Thomas Friedman, and Toronto *Globe and Mail* columnist Margaret Wente.

5. For a critique of the concept of cultural imperialism, see Tomlinson, *Cultural Imperialism*.

6. Ashcroft, Griffiths, and Tiffen, *Key Concepts in Postcolonial Studies*, 112–13.

7. Eagleton, *The Ideology of the Aesthetic*, 1.

8. On economics, see Hirst and Thompson, *Globalization in Question*; Burtless et al., *Globaphobic: Confronting Fears about Open Trade*; and Therborn, "Into the 21st Century: The New Parameters of Global Politics."

    On the long history of global migration and intercultural communities, see Bernal, *Black Athena* and McNeill, *Plagues and Peoples*. For a discussion of transformations in political modernity, see Hardt and Negri, *Empire* and Taylor, *Modernities*. Finally, Mattelart has emphasized recently the long-term development of that most important figure in the narrative of globalization: communications technologies. All of these works are, of course, examples drawn from a formidable body of texts and debates concerning claims about globalization's originality. Mattelart, *Networking the World, 1794–2000*.

9. Clausen, "Nostalgia, Freedom, and the End of Cultures," 234.
10. Appadurai, *Modernity at Large*, 32.
11. Appadurai, *Modernity at Large*, 52.
12. Grossberg, "Speculations and Articulations of Globalization"; Jameson, "Culture and Finance Capital," in *The Cultural Turn*.
13. Tomlinson, *Globalization and Culture*, 24.
14. Readings, *The University in Ruins*; Miyoshi, "Ivory Tower in Escrow"; Poovey, "The Twenty-First Century University: What Price Economic Viability?"
15. Cultural Industries Sectoral Advisory Group on International Trade (Canada), *Canadian Culture in a Global World*, 1.
16. Bourdieu, *Distinction*; Eagleton, *The Ideology of the Aesthetic*.
17. Balibar and Macherey, "On Literature as an Ideological Form: Some Marxist Propositions," 46.
18. Lloyd and Thomas, *Culture and the State*, 1.
19. Lloyd and Thomas, *Culture and the State*, 2.
20. Readings, *The University in Ruins*, 12.
21. Readings, *The University in Ruins*, 9.
22. Scarry, *On Beauty and Being Just*; Steiner, *Venus in Exile: The Rejection of Beauty in Twentieth-Century Art*; Elkins, *Pictures and Tears: People Who Have Cried in Front of Paintings*.
23. Miller, *Boxed In: The Culture of TV*, 8.
24. Miller, *Boxed In*, 6.
25. Benjamin, "The Work of Art in the Age of Mechanical Reproduction," 218; Bourdieu, *The Rules of Art*.
26. Marcuse, *Negations: Essays in Critical Theory*, 94–95.
27. Readings, *The University in Ruins*, 17.
28. Schor, "Towards a New Politics of Consumption."
29. Canclini, *Consumers and Citizens*, 39.
30. Canclini, *Globalization and Multicultural Conflicts*, 5.
31. Hansen, "The Mass Production of the Senses"; Buck-Morss, *Dreamworld and Catastrophe*.
32. Bull, "Between the Cultures of Capital," 102.
33. Bull, "Between the Cultures of Capital," 100.
34. Hardt and Negri, *Empire*.
35. Bourdieu, *Acts of Resistance*; Bourdieu, *The Rules of Art*, 339–44.

CHAPTER 3

Globalization, Postmodernism, and Literary Criticism (2006)

One can refute Hegel (perhaps even St. Paul) but not the Song of Sixpence.

—Northrop Frye, Conclusion to the *Literary History of Canada*

Globalization and Literary Studies

What possibilities does globalization open up for literary studies, and more specifically, for our understanding of the politics of the literary today? To put this another way: is it possible to still imagine a social function for literary studies in an era dominated by visual spectacle, the triumph of the private, and the apparent dissolution of the public sphere? Is there anything like a position of (relative) autonomy from which critics might reflect on the circumstances we have the misfortune to inhabit?

To speak of the opening up of new possibilities and even new political functions for literature and literary criticism today might seem quixotic at best: a tilting against the windmills of a radically transformed society that no longer has much use for the written word—or for culture more generally, for that matter. But if we attend carefully to globalization and consider how the practices of literature and literary criticism figure into the contemporary social and political landscape, it seems to me that some unexpected political possibilities emerge. While globalization signals the beginning of many new processes, those of us concerned with language, culture, and politics have often come to take it only as the name for the end of things: the end of democracy, of unmediated experience, of the public sphere, of the experiment (warts and all) called the Enlightenment, and, effectively, of poetry and literature, too. I want to argue that both literature and literary criticism have an essential political role to play in the era of globalization, even if they do so in transformed and difficult circumstances.

Integral to literary studies is the view that the "real" is always metaphorical in nature. All of our epistemologies, however secure and self-satisfied they might be in their ultimate veracity, are constituted by the appearance of the "real" in language: it is only by passing through metaphor that what is "outside" of language can become linguistic and thus intelligible at all. What better practice to challenge the self-certainties of the narratives of globalization—which function in part by denying their core metaphoricity—than literary theory and criticism? To grasp how and why the literary might provide the conditions for mutinous metaphors against the dominant ones articulated in the discourse of globalization, it is necessary first to describe (yet again) what globalization is (and is not) and how literature and the study of culture fits (or does not fit) into it; and so it is here that I begin.

Globalization Is Not Postmodernism

At the core of Karl Marx's investigation of the operations of capitalism is a sometimes forgotten critique of scholarly methodology. The political economists of his time mistook the *dramatis personae* of the modern economy—owners and workers—as *a priori* ontological categories, rather than as social positions that come into existence only as the result of a specific course of historical development. This methodological "failure" describes, of course, a more general process of reification that takes place throughout much of contemporary social reality and at many levels: our own creations take on the character of "natural," pre-ordained reality in a way that obscures the quotidian character of their invention. Marx's point goes beyond simply criticizing method. For one of the singular inventions of capitalism is the commodity form, which itself ceaselessly, on an ongoing and daily basis, *re-reifies* existing social relations. "The commodity," Marx writes, "reflects the social characteristics of men's own labor as objective characteristics of the products themselves, as the socio-natural properties of these things."[1] The commodity, one might say, acts as an objective reifying force that extends beyond the ideologies of capitalists and capitalism: we *live* this reification, whether we believe the larger social script in which it is embedded or not.

It should come as no surprise that "globalization" plays an important role in this ongoing narrative of capitalist reification. Just as surely as political economy for Marx, globalization hides reality from us even as it proposes to explain it. Just how does it do so? At first blush, the promise of the term "globalization" is that it offers us a way to comprehend a set of massive changes (clustered around the economic and social impact of new communications technologies and the almost unfettered reign of capital across the Earth) that have radically redefined contemporary experience. These changes cut across spheres of social experience *and* areas of scholarly analysis that were imagined previously to be separate (i.e., the economic, the cultural, the social, the political, and so on). And, confusingly, "globalization" names at one and the same time both the empirical and theoretical novelty of the processes most commonly associated with it. It names both a new reality and the new concept (or set of concepts) needed to make some sense of this reality. It is not surprising that this double role has made it an inherently unstable and amorphous concept, "used in so many different contexts, by so many different people, for so many different purposes that it is difficult to ascertain what is at stake in . . . globalization, what function the term serves, and what effects it has for contemporary theory and politics."[2] The immense debates that have ranged over what globalization "is" and what phenomena should (and should not) be included within it, the question of what the "time" of globalization might be (is it post-1989? the arrival of Columbus in the New World? the explosion of cross-regional trading in the eleventh century?), the issue of the politics of globalization and the possibilities of alternate globalizations to this one—all draw attention to the fact that the empirical realities that the term is meant to capture can potentially be arranged and re-arranged in very different and even contradictory ways. In other words, although globalization is at one level "real" and has "real" effects, it is also decisively and importantly rhetorical, metaphoric, and even fictional; it is a reality given a narrative shape and logic, and in a number of different and irreconcilable ways. But right away, one can also see that as soon as the idea of concept as metaphor—concept as not the thing itself (how could it be otherwise?) but as necessarily a substitution meant to produce an identity—is introduced, the real begins to fade away. What we take as the "real" of globalization necessarily comes mediated by the apparatus of numerous concepts strung together in an effort to grasp the fundamental character of the contemporary.

This characterization of globalization—as an amorphous term for the present; as an analytically suggestive and yet confusing concept that binds epistemology and ontology together; as an impossible yet compelling idea that names the logic organizing all experience; as a term that is potentially all things to all people and can be bent to multiple purposes—makes it sound like the successor to another concept that was intended to do similar kinds of work: postmodernism. Indeed, it is hard to avoid the idea that "globalization" carries out the periodizing task once assigned to postmodernism, naming the character and dynamics of the contemporary moment, if with far more attention paid to the material realities, struggles, and conflicts of contemporary reality on a worldwide scale. Globalization can thus appear to be a new and improved version of postmodernism, but one for which the issues of (for instance) the legacies of imperialisms past and present play a constitutive (instead of ancillary) role.

But as soon as this connection is ventured, it is clear that globalization is far from a replacement term for postmodernism. The differences between the two terms are instructive, especially with respect to the situation of literature and criticism at the present time. The postmodern was first and foremost an aesthetic category, used to describe architectural styles, artistic movements, and literary strategies,[3] before ever becoming the name for the general epistemic or ontological condition of Western societies—the "postmodern condition" that Jean-Francois Lyotard detected in his review of Quebec's educational system.[4] Criticisms of postmodernism focused on the adequacy of the term as an aesthetic descriptor (was postmodern fiction not really just more modernist fiction?), on its overreaching ambition at global applicability (was the "post" in "postmodernism" really the same as the one in "postcolonialism"?), or on the fact that there was far too little attention paid to the historical "conditions of possibility" of the emergence of the aesthetic and experiential facets of the postmodern. In short, this lack of attention hid the fact that postmodern style represented something more primary: the cultural logic of late capitalism.[5]

Whatever else one might want to say about globalization, it is clear that the term has little relation to aesthetics, or indeed, even to culture, in the way that postmodernism does. It is meaningless to insist on a global style or global form in architecture, art, or literature. There is no "globalist" literature in the way that one could have argued that there was a postmodernist one, nor a globalist architecture as there was (and still is) a postmodern one, even if there are global architects (such as Rem Koolhaas, Frank Gehry, or Zaha Hadid) and a global corporate vernacular in (say) airport or office tower design. This lack of relation to culture can be seen in the fact that we lack even the adjective for such a category—"global" literature being something very different from postmodern writing, without the immediate implications for form or style raised by the latter category. "World cinema" similarly names a moment rather than a style, though here perhaps one could argue that there has been a broad bifurcation of film into the cinema of the culture industry and the products of a new, globally-dispersed avant-garde (Hou Hsiao-hsien, Emir Kusturica, Agnès Varda, etc.); both can claim the title of "world cinema," if for wildly different reasons. "World poetry" names not even a moment in this sense, but simply the poetry of the whole world, samples of which we might expect to find collected in an anthology or reader of the kind that is constructed to be attentive to the differences of nation, region, and locality. The aesthetic may not have disappeared. But the category "global" as a periodizing marker doesn't address it, as if the ideological struggles and claims once named by the aesthetic and pursued by various avant-gardes have for some reason been rendered moot and beside the point.

If postmodernism comes to our attention through various formal innovations that prompt us to consider symptomatically what is going on in the world to generate these forms, globalization seems to invert this relationship. It places emphasis on the restructuring of relations of politics and power, the re-scaling of economic production from the national to the transnational, on the light speed operations of finance capital, and the societal impacts of the explosive spread of information technologies. With globalization, we thus seem to have suspended what was central to debates and discussions of postmodernism—the category of representation. Indeed, the contemporary reality named by globalization is meant to be immediately legible in the forces and relationships that are always already understood to be primary to it and to fundamentally constitute it (e.g., transnational economics, bolstered by the changing character of the state, and so on). What the comparison between postmodernism and globalization highlights is that there is not only no unique formal relationship between contemporary cultural production and the cultural-political-social-economic dominant named by globalization. In addition, there is apparently less reason to look to culture to make sense of the shape and character of this dominant, which apparently can explain itself, and which views culture as little more than a name for just one of the many aspects of commodity production and exchange today. Put another way, globalization seems to have transformed culture on the one hand into mere entertainment whose significance lies only in its exchangeability. On the other, it shifts it to refer to a set of archaic cultural practices that of necessity have little to say about the skylines of Shanghai's Pudong district or the favelas of Rio, other than to render an increasingly mute complaint about a world that has passed it by. If globalization is the postmodern come to self-recognition, it appears in the process to have transformed culture into mere epiphenomenon and to have rendered cultural criticism in turn into a practice now in search of an object. This state is evident especially as one of criticism's older political functions—making visible the signs and symptoms of the social as expressed in cultural forms—has been eclipsed by history itself.

This analysis might suggest that anxieties about the decline of (a certain vision of) culture in the era of globalization are in fact justified. But there is also another crucial difference between globalization and postmodernism that needs to be pointed to first, which will begin to turn us back to the question of the activity of literature and literary criticism in relation to globalization—and to the productivity of metaphor in relation to globalization as well. Postmodernism was never a public concept in the way that globalization has turned out to be. The postmodern never made anything more than a tentative leap from universities to the pages of broadsheets, appearing only occasionally in an article on the design of a new skyscraper or in sweeping dismissals of the perceived decadence of the contemporary humanities. It is a concept in decline, used these days mainly as a term for strange and incoherent phenomena or forms of social instability. By contrast, globalization is argued for by the World Bank, named in the business plans of Fortune 500 companies, and on the lips of politicians across the globe; it constitutes official state policy and is the object of activist dissent: the Zapatistas did not rise up against postmodernism, nor was it the preponderance of self-reflective, ironic literature in bookstores that brought anarchists into the streets of Genoa. There is clearly more at stake in the concept of globalization than there ever was with postmodernism. There is a politics that extends far beyond the establishment of aesthetic categories to the determination of the shape of the present and the future—including the role played by culture in this future. Even if both concepts function as periodizing terms for the present, globalization is about blood, soil, life, and death in ways that postmodernism could only ever pretend to be.

The public ambition of the concept of globalization makes it clear that there are two broad uses of this concept that need to be separated. Significantly, the confusions over the exact meaning and significance of globalization that have characterized much academic discussion have *not* in fact cropped up in the constitution of globalization's public persona. Far from it. The wide-ranging debate in the academy over the precise meaning of globalization might point to the fact that it is a concept open to re-narration and re-metaphorization, thereby keeping focus, too, on the unstable relationship between the realities the term names and its heuristic role in grappling with this reality. Like any concept, it is not equivalent to reality, but a way of producing some meaningful interpretive order out of the chaos of experience. Against this status, however, one must consider the function of the widespread public consensus that has developed on what globalization means. This is globalization in its most familiar garb: the name for a process that (in the last instance) is understood as economic at its core. Globalization here is about accelerated trade and finance on a global scale, with everything else measured in reference to these dimensions. Although one can have normative disagreements about the outcome and impact of these economic forces (does it "lift all boats," bringing prosperity to everyone? does it merely restore the power of economic elites after a brief interval of Keynesianism?), what the public discourse on globalization insists on is, first, the basic, immutable objectivity of these economic processes, and second, that these processes now lie at the core of human experience, whether one likes it or not.

It is in this way that the discourse of globalization carries out what has to be seen as its major function: to transform contingent social relations into immutable facts of history. It carries out this reifying function in a novel way. Unlike the categories of the political econo-mists of Marx's time, globalization insists not on the permanence of social classes, but on the coming into being of *new* social relations, technologies, and economic relationships. The overall effect is the same, however. Old-style political economy reified capitalism by insisting that existing social relations would extend indefinitely and unalterably into the future based on their origins in the very nature of things. New-style globalization also makes a claim on the inevitability of capitalism and the persistence of the present into the future. However, its necessary imbrication with the "new"—globalization always being the name for something distinctly different from what came before it—means that it cannot so easily appeal to nature or ontology to insist on the unchang-ing character of the future. Rather, borrowing a page from Marxism, globalization offers a narrative of the historical development of social forces over time, the slow (now accelerating) transformation of indi-viduals and societies from the inchoate mess of competing and warring nationalisms to a full-fledged global-liberal-capitalist civilization. Thus famously does Francis Fukuyama appropriate the movement of the Hegelian dialectic to capitalist ends.[6] He argues that the lack of alter-natives to capitalism signaled by the collapse of communism coincides with the "end of history" as such: there will only be capitalism from now on, and, of course, it will be everywhere, on a global scale. The erasure of the distinction between globalization as a conceptual apparatus and the name for contemporary reality as such is hardly an accident—or at least no more so than the categories of classical political economy. It is, rather, a political project through and through, meant (in the terms that I have outlined here) to deliberately confuse the potential analytic functions of the concept of "globalization" with an affirmation of un-changing reality of global capitalism as both "what is" and "what will be." In changing circumstances which *have* opened up new realities and political possibilities, the public face of globalization aims not only to keep capitalism at the center of things, but to clear the field of all possible challenges and objections.

Some clarification is in order here. I have claimed that globalization is a political project, which suggests some organizing force or set of actors or agents behind the scenes pulling the levers of state and econ-omy in order to shape the world into a desired state. This claim would make globalization a strictly ideological concept, a knowing sleight of hand by which the Grand Inquisitors of Davos pull the wool over the world's eyes. It would be naïve as well as empirically incorrect to deny that actors in industry and the state have actively participated in the reconstitution of relations between state and capital on a global scale for their own benefit, with consequences ranging from the release of public assets to the market at fire sale rates, to the increasingly precari-ous state of global labor markets.[7] At the same time, there is a tendency by many critics to ascribe too much insight and control over the system of neoliberal globalization to specific individuals (CEOs, government leaders, etc.) or institutional elements (government agencies, WTO, IMF, etc.). It is as if to suggest that these actors view globalization from the outside and with a clarity that allows for the perfect decision to be made in every case.

The politics of our global era do not permit an easy reliance on a vision of the social order in which change can be achieved by cutting off the head of the king. Globalization as an ideological discourse (in the way I have described it) appears within an already entrenched social and political system, which is the product of the dynamics and technics of modernity's structuring of the social order and the production of subjectivities. This modernity is one whose logics, it has to be added, extended across the ideological divide of the Cold War: modernization and Taylorization represented the future for the Soviets and the West alike. The fundamental drive of the system as a whole continues to lie in the core imperative of capitalism: the unlimited accumulation of capital by formally peaceful means.[8] As Michael Hardt and Antonio Negri argue, the tension that exists within this social fantasy—endless accumulation without strife—has been dissipated historically through the availability of an "outside" to the system of capital where surpluses can be actualized, thus avoiding the potential social trauma of over-production.[9] The moment when capital finally finds itself victoriously spread across the globe—its extensivity confirming its supposed superiority as a social and economic system—is also a moment when its contradictions, inhumanity, and fundamental absurdity become increasingly evident, especially as processes of "accumulation by dispossession" accelerate.[10] As the collective Retort points out, "insofar as the spectacle of social order presents itself now as a constant image-flow of contentment, obedience, enterprise, and uniformity, it is, equally constantly, guaranteed by the exercise of state power. Necessarily so, since contentment, obedience, enterprise, and uniformity involve the suppression of their opposites, which the actual structure and texture of everyday life reproduce—and intensify—just as fast as the spectacle assures us they are things of the past."[11]

In this context, both ideology and state intervention reappear as necessary to maintain order and stability. The public discourse of globalization engages in the effort to secure the existing social order at all costs, but not only because of the obvious benefits it provides to some. There is a systemic effect at work, which comes out of deep, intensive social commitments to order, expertise, technology, progress, consumption, and capital. Margaret Thatcher's turn to the ideas of F. A. Hayek, Milton Friedman, and others, originates not as a strictly ideological move. It is one occasioned by the need to resolve seemingly intractable economic problems within the existing framework of liberal democracy. Although the championing of markets, private property, and entrepreneurial energies may have pushed the state toward the market and away from social welfare, commitments to these ideals were hardly external to the modern state to begin with. All power here is on the side of modernity. In the absence of compelling or convincing alternative political narratives, the social chaos engendered by neoliberalism all the more powerfully confirms its necessity. Existing systems alone appear to have the capacity to manage the radical economic and social change that has produced the economic instability and social precariousness in which we all live.

Globalization and Literary Criticism

How does this account of globalization open up new possibilities for literature and literary criticism? Perhaps the major response to globalization within literary studies has been to redefine its practices in light of a world of transnational connections and communications. Globalization has often been interpreted as signaling the end of the nation-state and of the parochialisms of national culture. Waking up to the limits of its own reliance on the nation as a key organizing principle, literary studies and poetics have thus come to insist on the need to take into account the global character of literary production, influence, and dissemination. Much of contemporary literary studies has focused correspondingly on the transfer and movement of culture: its shift from one place to another, its newfound mobility, and the challenges of its extraction, de-contextualization, and re-contextualization at new sites. At one level, this encounter of criticism with "globalization" has simply required the extension or elaboration of existing discourses and concepts, such as diaspora, cosmopolitanism, the politics and poetics of the "Other," and the language of postcolonial studies in general. For many critics, literary criticism was already moving toward globalization in any case, or was even there in advance, as suggested by accounts stressing the existence of global literary relations long before the present moment.[12]

There have been other developments as well. There has once again been serious attention to the politics of translation and renewed focus on the institutional politics of criticism, especially the global dominance of theory and cultural criticism by Western discourses.[13] There have also been new sociologically-inspired "mapping" projects that have sought to explore how literary and cultural forms have developed and spread across the space of the globe.[14] Finally, criticism has taken up an investigation of new literary works whose content, at least, criticizes and explores the tensions and traumas produced by globalization—a potentially huge set of works given the fact that globalization is often taken to be coincident with contemporary geo-politics as such. There have been rich critical discoveries in every one of these attempts to take up in literature and criticism the challenges—real or imagined—posed by globalization.

However productive and interesting such analyses are, there is nevertheless a way in which they are all too willing to take globalization at face value. They acquiesce to the character and priority of capital's own transnational logics and movements, instead of questioning and assessing more carefully the narrative that underlies them. The critical agenda is thus set by the operations of globalization *qua* global capital. The need for criticism to concentrate its own energies on movement and border-crossings, while not entirely misplaced, comes across as rearguard maneuvers to catch up with phenomena that have already taken place at some other more meaningful or important level. In this anxious attempt to claim the terrain of the global and the transnational for culture and criticism, the minimized role of culture within the narrative of globalization that emerges out of the comparison of globalization with postmodernism is troublingly reaffirmed, even if this affirmation is not the intent of these various and varied new approaches to culture in the era of globalization.

This is not to say that the approaches to globalization described above are without impact or value. It is simply to call attention to the fact that the project called globalization demands other responses that address directly its rhetorical and fictional character, and in particular, the ideological attempt to seal off the future through the assertion of a present that cannot be gainsaid. At one level, such a response would simply be to remind us insistently of the fiction that is the public face of globalization. It calls attention to and exposes the endless employment of rhetoric in the struggle over the public's perception of the significance and meaning of the actions of businesses and governments, peoples and publics in shaping the present for the future, and indeed, in shaping what constitutes "possibility" itself. What better practice to do this than literary criticism, which is characterized by nothing other than its attention to the powerful uses (and abuses) of language in shaping and mediating our encounter with the world? The consistent anthropomorphisms applied to globalization, which make globalization into a beast that penetrates markets, speeds up time, breaks boundaries, and changes the world seemingly independently of human involvement is one of the key issues that criticism can bring to the fore.

This is just one possibility, and one which still seems to leave the literary in the dust of globalization by turning literature and literary criticism into a broader form of cultural criticism; its continued utility is justified only by its usefulness as a tool against ideology. The object of literary studies in this case would be the tropes and turns of language used explicitly to shape public perception: "axis of evil," "weapons of mass destruction," "democracy," "progress," and even "development," "empowerment," and the like.[15] The political possibilities of literature and criticism today are in any case larger and more general than these terms, if also perhaps less satisfactorily and explicitly definable, and, unfortunately, more troubled and difficult as well. I have introduced two senses of globalization: one that remains open to debate and re-narrativization, even about so fundamental an issue as "when" globalization might be; and another that seems to know definitively when (now) and what (global trade) globalization is. The second globalization aims to undo and even to eliminate the contradictions and confusions opened up by the first, in order to reassert capitalism's ontological legitimacy. The political possibilities that globalization opens up for the literary can be grasped only by asking the question of why capitalism needs the new rhetoric of "globalization" at this time. Why does the lumbering beast of capital have to be re-described and given perhaps even greater autonomy than it possesses in its most metaphorically potent guise as the "invisible hand"? Do not the old categories of political economy continue to assert their mystificatory role in the ways that they have for so long?

The negative answer to this last question is pointed to in the very instability of the concept of globalization. Its claim to articulate uniquely the new and the future leaves it open to endless doubts and questions that require its ideological dimensions to be affirmed anew over and over again[16]—not least as a result of the "suppression of opposites" described above by Retort. Globalization is breathlessly confident, a master narrative that demands that all other concepts, ideas, and practices be redefined in relation to it. And yet, the insistence of globalization narratives on the absolute priority of the economic also interrupts its legitimacy at the moment it imagines itself as most forcefully asserting it.

Critical Imaginings

In the colonization of the globe by capital, and the simultaneously geographic spread of communication technologies and cultural forms of all kinds, we might imagine that the reign of commodity fetishism, for instance, is affirmed as never before. But as capital reaches the limits of the globe, there is another story emerging that shakes its hold over the future. If the globalization of production has necessitated new narratives of the "good" of trade liberalization—the "good" of capital—it is because the complex, dispersed modes of contemporary production have not hidden away the social realities of production in the absent corners of the globe, but have rather drawn ever more attention to the social relations embedded in commodities. In *Capital*, Marx famously writes that "so soon as [a table] steps forth as a commodity, it is changed into something transcendent. It not only stands with its feet on the ground, but, in relation to all other commodities, it stands on its head, and evolves out of its wooden brain grotesque ideas."[17] But what tables today dare to evolve out of their wooden brains grotesque ideas or dance of their own free will? They must instead give an account of their productive parentage: from where did they come? How and by whom were they made? (By child laborers? By well-paid unionized workers?) For what purpose? Under what conditions? (In sweat shops? On industrial farms? In third-world tax havens?) And at what cost to that ultimate social limit, the environment? Though no less part of the system of exchange, the commodity today can no longer be depended on to buttress capitalism by shielding from view the social relations that create it. The response offered by the narrative of globalization is not to hide these social relations, but to claim first their inevitability, and then to provide a utopic future-oriented claim about a coming global community in which the traumas of the present will be resolved in the fluid shuttling of freely-traded goods around the world.

The utopia offered by the dominant narrative of globalization is one that has to be rejected, perhaps along with the concept itself, which has become so deeply associated with the current drive and desire of capital as to make it now almost impossible to wrest anything conceptually productive from it. The focus should instead be on the production of new concept-metaphors that might open up politically efficacious re-narrativizations of the present with the aim of creating new visions of the future. For all its ubiquity and hegemonic thrust, the instability of the concept of globalization presents an opportunity to do so; and so, far from being sidelined in globalization, there is an opening for creative critical thinking of all kinds to intervene and generate alternatives. It is here that literary and cultural production and literary criticism have roles to play: not only to shock us into recognition of reality through ideological critique, but also to spark the imagination so that we can see possibility in a world with apparently few escape hatches.

Why concept-*metaphor*? At its most basic level, metaphor involves the production of identity through substitution in a manner that opens up new and unexpected relationships and ideas. Metaphor is fundamental to literary language; it is what distinguishes it from mere reportage, nonfiction, or journalism (which is not to say that these genres do not employ metaphor as well). The phenomenological chaos that those concepts circulated between state and institutional social science are meant to tame or foreclose is the very medium of literary and cultural narrative—what they puzzle over and tarry with. While elements of the discourse of globalization may employ metaphor, globalization as such is *anti*-metaphoric. Even as it appeals to innovation and creativity for its increasingly immaterial, informational economy, it nonetheless demands a resolution or adjournment of time in order to control and manage the newness thus brought into life. This is no doubt why, as I have argued earlier, the aesthetic has disappeared from globalization; if "culture" shows up at all, it is in the guise of a commodity that contributes to economic vitality (as in Richard Florida's "creative class") or as a form whose main purpose is to ameliorate social problems through state cultural programs and national cultural policy.[18] Through metaphor, on the contrary, temporality is subjected to interrogation, and dead objects and concepts are brought back to life through the evocation of impossible identifications. It is in this way that newness comes into the world and the present is not all that remains.

For what is genuinely lacking today is the imaginative vocabulary and narrative resources through which it might not only be possible to challenge the dominant narrative of globalization, but to articulate alternative modes of understanding those processes that have come to shape the present—and the future. This need is often narrowly imagined as a political lack, the absence of a big idea to take the place of state socialism after the collapse of the Soviet Union and the colonization of the Western left by disastrous "third way" political approaches. The imaginative resources that are needed to shape a new future are, however, necessarily broader—or at least, a new political vision is impossible without a revived poetics of social and cultural experience as well. This evocation of imagination in relation to poetics and the politics of globalization can be read in the wrong way: at best, as an appeal to Arjun Appadurai's still shaky use of "imagination" in his influential *Modernity at Large* (1996); at worst, as a Romantic, idealist faith in the autonomous origin of ideas and their power to shape reality.[19] What I have in mind is neither of these, but rather Peter Hitchcock's use of "imagination as process" in his account of the promise of a theoretical maneuver that would be able to seize upon the conceptual openings that "globalization" has generated within capital itself. He writes that

> [w]hile there are many ways to think of the globe there is yet no convincing sense of imagining difference globally. The question of persuasiveness is vital, because at this time the globalism most prevalent and the one that is busily being the most persuasive is global capitalism. To pose culture alone as a decisive blow to global modes of economic exploitation is idealist in the extreme . . . Yet, because such exploitation depends upon a rationale, a rhetoric of globalism if you will, so culture may intervene in the codes of that imaginary, deploying imagination itself as a positive force for alternative modes of Being and being conscious in the world.[20]

There is a great deal that can be said here about the possibilities and limits of literature and literary criticism in reference to the imagination and persuasiveness. On the one hand, it is meaningless to assert that literature in general produces, through narrative and through metaphor, social visions other than the ones we work through in daily life. The kind of genre literature that comprises most of the market for literary texts reinforces the dynamics and logics of capitalism. Or does it? Even in such cases, the need to reproduce the entire world in fictional form recreates, whether implicitly or explicitly, the tensions and contradictions between the experience of the world and the discourses meant to describe this experience. In other cases, from Jamaica Kincaid's *A Small Place* to Mahasweta Devi's *Imaginary Maps*, or from Paulo Lins's *City of God* to Peter Watts's *Rifters* trilogy (which explores a capitalism that persists into the future despite its intense contradictions), the aim is precisely to give flesh to the abstractions of globalization and to highlight the contradictions of neoliberalism. The point here is to insist on the importance of these imaginings, drenched in the metaphoric, as a counterweight to those discourses of globalization that claim to have already put everything in its place, including literature and culture more generally. What is more difficult to assert and to argue for is the significance or importance of this or that specific text, their persuasiveness, or their impact on imagination and the generation of "alternative modes of Being." In his exploration of the increasing use of "culture as resource" today, George Yúdice writes that "the role of culture has expanded in an unprecedented way into the political and economic at the same time that conventional notions of culture largely have been emptied out."[21] If literary texts and critical approaches to them do not constitute a program to upend or overcome the deprivations and limits of globalization, at a minimum they engage in a refusal of the contemporary prohibition on metaphor and its imaginative possibilities.

Rather than give a determinate account of the how and why of the ways in which culture can intervene into the imaginary, I want to leave this sense of imagination open and suggestive, and end by discussing briefly one more shift for aesthetics in general and literature in particular in relation to globalization. If we are to speak about the imaginary and its powers in the way Hitchcock does, we can do so today only in reference to an aesthetic that is very different from the one normally conceptualized. This is an aesthetic that no longer claims its potential political effect by being transcendent to the social, but by being fully immanent to it. A half-century or more of literary and cultural criticism has insisted that culture be viewed as part of the social whole—generated out of and in response to its contradictions, its certainties as well as its uncertainties, an exemplar of its division of labor and its use of symbolic forms to perpetuate class differences through the game of "distinction." For those invested in a literary or cultural politics premised on a vision of the autonomy of art and culture from social life, the demand to take into account the social character of the literary comes as a loss, as does the more general massification of culture, which seems to announce the draining of the energies of the poem, the novel, and the art work. Insofar as globalization has also been seen as announcing a "prodigious expansion of culture throughout the social real, to the point at which everything in our social life . . . can be said to have become 'cultural,'"[22] it, too, seems to suggest the general decline of the politics of culture. This is no doubt why globalization is construed as a threat to poetics: it is nothing less than mass culture writ large over the face of the globe.

But this is the wrong lesson to draw from the folding of the aesthetic into the social, or of the expansion of culture to encapsulate everything. In his assessment of the politics of avant-garde, Peter Bürger identifies the contradictory function of the concept of "autonomy" in the constitution of the aesthetic: it identifies the real separation of art from life, but covers over the social and historical origins of this separation in capitalist society. The aim of the historical avant-garde—and perhaps I could venture to say all artistic movements since Kant—is to reject the deadened rationality of capitalist society through the creation of "a new life praxis from a basis in art."[23] Bürger suggests that this had already happened by the middle of the twentieth century. Art had been integrated into life, but through the "false sublation" of the culture industry rather than through the avant-garde. In the process, he claims that what has been lost is the "free space within which alternatives to what exists become conceivable."[24] Yet to see the sublation of art into life through mass culture as "false" or as a "loss" requires the affirmation of the problematic autonomy of art from life produced by social divisions that we should be glad to see dissolved. That these divisions have not been dissolved by the culture industry, but have taken new forms, is clear; equally clear, however, should be the fact that the ability of culture to conceive alternatives, far from lost, has been diffused across the spectrum of cultural forms, which is why the imaginative capacity I am pointing to above can potentially come from anywhere. What an immanent aesthetic lacks that a transcendent one possessed in spades is that revolutionary spirit that animated nineteenth- and twentieth-century politics and culture, in which the right moment or perfect cultural object could—all on its own—shatter the ossified face of social reality. The writer or artist as vanguardist guardian of the good and the true is definitively over. But to this we can only say: good riddance. Welcome instead a politics and poetics that proceeds uncertainly, through half-measures and missteps, through intention and accident, through the dead nightmare of the residual and the conservative drag of hitherto existing reality on all change, in full view of the fact that nothing is accomplished easily or all at once, or in the absence of the collective energies of all of humanity, and through the imaginative possibilities of literature, yes, but other cultural forms, too.

A Final Word

Repudiation of the present cultural morass presupposes sufficient involvement in it to feel it itching in one's finger-tips, so to speak, but at the same time the strength, drawn from this involvement, to dismiss it.

—Theodor Adorno, *Minima Moralia*

So much for the function of culture at the present moment. But I cannot end without asking directly: what of the autonomy of the critic? The question of autonomy—the precise relationship of structure to agency (if, indeed, agency exists at all) in the field of culture—has been perhaps the central theme of literary and cultural criticism since 1968. Every major critic and theorist of the period, from Louis Althusser to Slavoj Žižek, has struggled to formulate how individual and/or collective agency functions in a period in which—for reasons of global scale as well as technological "advance"—human social life seems shaped and controlled in ever more extreme, invasive, and unstoppable ways. Thus we have the development of ideas such as Michel Foucault's biopower,

Pierre Bourdieu's *habitus,* and numerous others, which offer narratives of a structure that generate the very forms of (imagined) agency that might then want to resist the system. It is a classic chicken-and-egg problem, if one encumbered with more abstruse vocabulary and the weight of the whole history of philosophy.

If individual subjects are produced by the structures in which they live (and how could it be otherwise?), the very ability of critical thought to grasp these structures—to paint a picture of the circumstances in which we find ourselves, clearly and explicitly as if from the "outside"—must necessarily come into question. At the end of *Dialectic of Enlightenment* (1944), their savage and hugely influential account of the dead end of the Enlightenment in the parking lot of the cultural industries, Max Horkheimer and Theodor Adorno pause to reflect on the conditions that make possible their own critical vantage point. They claim that their critique can emerge only because they occupy an unusual position: the interstices between the end of classical German philosophy and the burgeoning of American mass culture.[25] It is only because the former is radically incommensurate with the latter that Horkheimer and Adorno could "hate [both of them] properly"[26] and in so doing introduce the mode of criticism that has come generally to be referred to as critical theory.

This is a very specific location, a critical vantage point created only as result of the trauma of World War II, which saw members of the Frankfurt School relocated to Los Angeles. But this structural misfit is perhaps more general than this example might suggest, extending to *all* literary and cultural criticism. Antonio Gramsci made a useful distinction between two kinds of intellectuals: traditional and organic. Organic intellectuals are those who work with ideas and thus give capital "homogeneity and an awareness of its function not only in the economic but also in the social and political fields."[27] For Gramsci, the exemplary organic intellectual for the contemporary age was the engineer; in our moment of finance capitalism and dreams of the "creative class," academics and cultural workers—all those who might be referred to as "content providers"[28]—are if not examples of model intellectuals, then certainly part of the mix of the "idea men" necessary for the functioning of the present.

At the same time, what distinguishes intellectuals in the humanities from their brethren in some other parts of the university (research scientists, engineers) is that they also occupy a position that is decidedly out of step with the times—a "residual" position that makes them both connected to, and yet separate from, the dominant ideas of the moment (Bourdieu's memorable phrase for this class was "the dominated fraction of the dominant class"). To put this point in Gramscian terms, literary and cultural critics are also "traditional" intellectuals—traditional understood here not as an evaluative term but as the name for a *structural* category that does not depend on the subject of analysis (Manet or multimedia, Flaubert or film). For Gramsci, traditional intellectuals are connected to one another across time. Since "traditional intellectuals experience through an *'esprit de corps'* their uninterrupted historical continuity and their special qualification, they put themselves forward as autonomous and independent of the dominant social group"[29]; the lack of synchrony with the present, sometimes experienced by intellectuals as a frustration, is in fact a source of their analytic strength. It is this simultaneous connection to the present and filiation to the past that creates a position in which a degree of autonomous thought—a critical vantage-point on the present—is made possible.

And it is perhaps at the moment of globalization that this vantage point emerges with special force. As I suggested above, culture in globalization is simultaneously imagined to be everywhere and yet treated as inconsequential as a result of this very ubiquity. The attempt by the narrative of globalization to write cultural criticism out of the picture is a sign of the force and importance of this practice. Slavoj Žižek asks: "Does not the critique of ideology involve a privileged place, somehow exempted from the turmoils of social life, which enables some subject-agent to perceive the very hidden mechanism that regulates social visibility and non-visibility? Is not the claim that we can accede to this place the most obvious case of ideology?"[30] Put this way, it is essential to be vigilant about the simple belief that the critic stands above and outside of social life in a way that allows her to understand the mechanisms that produce reality. But it would be equally ideological to acquiesce to a global narrative that has tried ceaselessly to transform this genuine and necessary critical suspicion of a pure critical autonomy into a belief that literary and cultural critics cannot do anything at all.

NOTES

1. Marx, *Capital*, vol. 1, 165.
2. Douglas Kellner, "Globalization and the Postmodern Turn," 23.
3. Anderson, *The Origins of Postmodernity*.
4. Lyotard, *The Postmodern Condition*.
5. Jameson, *Postmodernism*, 1–54.
6. Fukuyama, *The End of History and the Last Man*.
7. Arrighi, "Hegemony Unravelling—I"; Comaroff and Comaroff, "Millennial Capitalism: First Thoughts on a Second Coming"; Harvey, *A Brief History of Neoliberalism*; and Harvey, *The New Imperialism*.
8. Budgen, "A New 'Spirit of Capitalism.'"
9. Hardt and Negri, *Empire*, 221–39.
10. Harvey, *The New Imperialism*, 137–82.
11. Retort, "An Exchange on *Afflicted Powers: Capital and Spectacle in a New Age of War*," 8.
12. Greenblatt, "Racial Memory and Literary History."
13. Spivak, *Death of a Discipline*; Kumar, ed., *World Bank Literature*.
14. Casanova, *The World Republic of Letters*; Moretti, *Modern Epic*.
15. Cornwall and Brock, "What Do Buzzwords Do for Development Policy? A Critical Look at 'Participation,' 'Empowerment' and 'Poverty Reduction.'"
16. For two recent examples, see John Tierney, "The Idiots Abroad," *New York Times*, Nov. 8, 2005: A27, and "The New World," *Der Spiegel*, Special International Edition 7 (2005): 3.
17. Marx, *Capital*, vol. 1, 165.
18. Yúdice, *The Expediency of Culture*.
19. Appadurai, *Modernity at Large*.
20. Hitchcock, *Imaginary States*, 1.
21. Yúdice, *The Expediency of Culture*, 9.
22. Jameson, "Marxism and Postmodernism," in *The Cultural Turn*, 48.
23. Bürger, *Theory of the Avant-Garde*, 49.
24. Bürger, *Theory of the Avant-Garde*, 54.
25. Horkheimer and Adorno, *Dialectic of Enlightenment*.
26. Adorno, *Minima Moralia*, 52.
27. Gramsci, *Selections from The Prison Notebooks*, 5.
28. O'Brien and Szeman, *Content Providers of the World Unite!*
29. Gramsci, *Selections from The Prison Notebooks*, 7.
30. Žižek, "Introduction: The Spectre of Ideology," 5.

CHAPTER 4

System Failure: Oil, Futurity, and the Anticipation of Disaster (2007)

Siri sipped at her coffee. "I would have thought that your Hegemony was far beyond a petroleum economy."

I laughed. . . . "Nobody gets beyond a petroleum economy. Not while there's petroleum there."

—Dan Simmons, *Hyperion*

And lurking behind any possible reconfiguration of world politics would be questions of access to energy and to water, in a world beset by ecological dilemmas and potentially producing vastly more than existing capacities of capitalist accumulation. Here could be the most explosive issues of all, for which no geopolitical manoeuvring or reshuffling offers any solution.

—Immanuel Wallerstein, "The Curve of American Power"

The way one establishes epochs or defines historical periods inevitably shapes how one imagines the direction the future will take. And so it is with the dominant periodization of the history of capital, which has been organized primarily around moments of hegemonic economic imperium: Dutch mercantilism, British imperialism, US transnationalism. All the effort in reading the tea leaves of contemporary capitalism is thus directed at determining when the current hegemonic formation will collapse and which new one (or ones) will come in its stead. According to Giovanni Arrighi, David Harvey, Immanuel Wallerstein, and others, the US moment is at an end; the new hegemonic formation will emerge only after a turbulent and violent interregnum that is already upon us, even if we do not yet recognize it. Through it all, it seems, capitalism emerges largely unscathed: different in content, perhaps, and no doubt occupying a different space on the globe, but essentially the same in form—a system organized around limitless accumulation, at whatever social cost.

What if we were to think about the history of capital not exclusively in geopolitical terms, but in terms of the forms of energy available to it at any given historical moment? So steam capitalism in 1765 creates the conditions for the first great subsumption of agricultural labor into urban factories (a process of proletarianization that is only now coming to a completion), followed by the advent of oil capitalism in 1859 (with its discovery in Titusville, Pennsylvania), which enabled powerful and forceful new modalities of capitalist reproduction and expansion. From oil flows capitalism as we still know it: the birth of the first giant multinationals—Standard Oil (whose component elements still persist in Exxon Mobil, Texaco, and British Petroleum), DuPont, and the Big Three automobile makers; the defining social system of private transportation—cars, air travel, freeways, and with these, suburbs, "white flight," malls, inner-city ghettoization, and so on; and the environmental and labor costs that come with access to a huge range of relatively inexpensive consumer goods, most of which contain some product of the petrochemical industry (plastics, artificial fibers, paints, etc.) and depend on the possibility of mass container shipping. No petroleum, no modern war machine, no global shipping industry, no communication revolution. Imagined in geopolitical terms, the future is one in which US hegemony gives way to (say) a Sino-Russian bloc or perhaps to some hydra-headed creature made up of economies that have passed from socialist caterpillar to capitalist butterfly (Brazil-India-China). But if we think of capital in terms of energy, what do the tea leaves tell us about what comes next? Wind capital? Solar capital? Biomass capital? A seemingly impossible conjunction of terms. Nuclear capital? Hydrogen capital? These are somehow more imaginable—even if the technological problems of the latter have yet to be worked out and the nuclear option would require a staggering and unprecedented investment in building new reactors, an expenditure that is not on the horizon anywhere.[1]

Oil capital seems to represent a stage that neither capital nor its opponents can think beyond. Oil and capital are linked inextricably, so much so that the looming demise of the petrochemical economy has come to constitute perhaps the biggest disaster that "we" collectively face. The success of capital is dependent on continuous expansion, which enables not only profit taking but investment in the reproduction of capital that is a necessary condition for its continuation on into the future. During the period of oil capital, this expansion and reproduction was fueled by cheap and readily available sources of oil, not least (until the early 1970s) in the United States itself. In "Critique of the Gotha Program," Karl Marx reminds us that "labor is not the source of all wealth. Nature is just as much a source of use values (and it is surely of such that material wealth consists!) as labor which is itself only the manifestation of a force of nature, human labor power."[2] The discourse of disaster around the end of oil recognizes that, unimaginably, at least one part of the use values originating in nature (the one that only seems to come for free) is on the verge of being exhausted. What happens to capital now that oil is (at best) likely to remain expensive or is (at worst) actually running out? What future can be imagined not only for oil capital, but for capital as such?

There seems to be general agreement that even if we have yet to reach "Hubbert's peak"—the point of maximum global oil supply prior to its downward decline to zero—the point at which we will is coming soon.[3] More optimistically (at least from one perspective), taking account of all possible sources of fossil fuels—tar sands, hard-to-access or currently off-limits sources of oil and gas (in the deep sea, in natural reserves, and in national parks), and especially coal—some economists and resource experts have estimated that the global economy could continue to be hydrocarbon based for 200 to 500 years, depending on the levels of growth in energy usage.[4] Whether oil is disappearing or in relative abundance, the recent triangulation of military adventurism, the demands of rapidly expanding developing economies as a result of globalization, and the cold hard facts of global warming and ecological catastrophe has led to a feverish explosion of discourses about the probable future of oil—and, to a lesser degree, of oil capitalism. It is the orientation of these discourses toward the "disaster" of the end of oil and the potential futures with which I will concern myself in this essay. Though the scientific veracity of the claims made on behalf of this or that narrative of the likely fate and future direction of oil capital is not incidental, it is also the case that the power of these narratives and the likelihood that one or another is adopted (to whatever degree and however incompletely) as a way of precluding the collapse of oil capital depend less on the judiciousness of the notoriously shaky predictions of petroleum geologists or traders in commodity futures than on the way in which they mobilize and intersect with existing social narratives of expertise, technology, progress, consumption, nature, and politics. In the case of such narratives, precise statistics and measurements hardly begin to capture the social anxieties, fears, and hopes embodied in discourses that try to imagine the shape of future social formations.

Is the end of oil a disaster? This depends, of course, on the perspective one has on the system in danger of collapse: capitalism. The disaster discourses of the end of oil are necessarily anticipatory, future-oriented ones—narratives put into play in the present in order to enable the imagined disaster at the end of oil to be averted through geopolitical strategy, rational planning, careful management of resources, the mobilization of technological and scientific energies, and so on. What is all too frequently absent from these quintessentially modern discourses is the shape and configuration of the political. Eco-dystopians and techno-utopians alike take the current configuration of the political and economic as given. Because of this, it seems impossible from these perspectives to envision a systemic revolution. This deficit within existing narratives of the end of oil should alert us to the largely unarticulated political possibilities that lurk within them. The task here, then, will be to critically assess existing "end of oil" narratives in order to consider their lessons for a left that has the difficult task of generating and articulating alternatives to oil capital. While the equation "blood for oil" effectively draws attention to one dimension of the geopolitics of oil, it leaves unaddressed how one conceptualizes energy demands for a human polity that is expected to grow to 9 billion by midcentury. Indeed, in celebrating the possibilities of the potentialities of South America's "Bolivarian Revolution" or the continued attractions of even latter-day Scandinavian social democracy (which, especially in the case of Norway, is fueled by oil), the left has seemed to resist thinking too deeply about the larger consequences of petroeconomies, of their sustainability as social and political models, and of what, if anything, comes after.

What might a left position on oil capital—and its aftermath—look like? There are three dominant narratives circulating today concerning what is to be done about the disaster of oil: strategic realism, techno-utopianism, and eco-apocalypse. In what follows, I take each one up in turn in order to see what lessons they have to offer for the left, before concentrating on one of the most contentious recent confrontations with oil capitalism: the "Blood for Oil?" chapter of Retort's *Afflicted Powers*.[5]

Strategic Realism

It has become a given that contemporary geopolitical maneuvering is driven by access to goods and resources, chief among these being access to oil. In describing the actions and motivations of imperialist jockeying between the major powers at the turn of the last century, Lenin evokes the name Standard Oil, but only as one of the many capitalist monopolies that had established themselves by the beginning of the twentieth century: there is nothing to distinguish it from any of the others he lists, such as the Rhine-Westphalian Coal Syndicate, United States Steel, the Tobacco Trust, and so on.[6] Today, steel and cigarettes have receded, and oil has come to the fore as a prime factor guiding the political decision making and military actions of both advanced capitalist countries and developing ones. As Daniel Yergin notes, oil arrived on the geopolitical stage at the outset of World War I when Winston Churchill, First Lord of the Admiralty, decided to power Britain's navy by oil from Persia as opposed to coal from Wales—a shift designed to improve the speed of the navy, but at the expense of national energy security.[7] This founding equation between oil and military power has been consistently in force ever since. The political character of the Middle East in particular has been shaped throughout the past century by the military and political struggle of Britain, France, the United States, and other powers to secure access to a commodity essential to the smooth operation of their economies.[8]

The Advanced Energy Initiative (AEI), announced early in 2006 by President George W. Bush's administration, is intended to reduce dependency on foreign oil by promoting the clean use of oil, nuclear power, natural gas, and a variety of renewable resources.[9] "Let me put it bluntly. . . . We are too dependent on oil," Bush stated at a 2006 conference organized by the US Departments of Energy and Agriculture to promote the use of biofuels (such as ethanol) in support of the AEI.[10] This bluntness and the announced aim by the administration to support new forms of energy have little impact on the necessity, at the moment and for the foreseeable future, to do whatever it takes to keep oil flowing into the US economy. As the AEI notes, it is not just the US economy that requires oil, but countries such as China and India, which are consuming more oil and at an accelerating rate.[11] Even if the United States and other major consumers of oil (China, Japan, Russia, Germany, etc.) should manage to reduce consumption and develop alternative sources of energy, there is no question that it remains an essential commodity, growing in demand even as its supply decreases.

What I have termed *strategic realism* is a relatively common discourse around oil that derives from a strict realpolitik approach to energy. Those who employ it—and it is a discourse employed widely by government and the media alike—suspend or minimize concerns about the cumulative environmental disaster of oil or the fact that oil is disappearing altogether, and focus instead on the potential political and economic tensions that will inevitably arise as countries pursue their individual energy security in an era of scarcity.[12] What is of prime interest in strategic realism is engaging in the geopolitical maneuverings required to keep economies floating in oil. At the heart of strategic realism stands the blunt need for nations to protect themselves from energy disruptions by securing and maintaining steady and predictable access to oil.

These maneuverings around energy can and do take multiple forms, from military intervention intended to shore up existing "power interdependencies"[13] (due to the US invasion of Iraq, military intervention has come mistakenly to stand as the prime mode through which access to oil is secured) to economic agreements between states, and from the creation of new trade and security arrangements of mutual benefit to the big users of oil (looking down the road, the United States, China, and India)[14] to even the (largely) quixotic attempt to create energy independence by promoting the use of alternative fuels. What ties these various approaches together is an element so obvious that it appears hardly to need mentioning: the centrality of the nation-state itself in the calculations of oil accessibility and security. When it comes to the potential disaster of oil, in the discourse of strategic realism the figures, concepts, and protagonists that we have all come to love in the discourses of globalization—the withered nation, Colossus-like transnational corporations, the mixed sovereignty of empire—seem not only to fade to the background but to disappear altogether. Strategic realism is a discourse that makes the nation-state the central actor in the drama of the looming disaster of oil, an actor that engages in often brutal geopolitical calculations in order to secure the stability of national economies and communities. While oil is hardly divorced from the operations of global finance, its political value as a commodity is such that it is apparently not permitted to slosh autonomously through markets that we have been repeatedly told take little note of borders today: the state must be present in order to ensure that every day the right amount of oil flows in the right direction.

Discussions of the strategic calculations at work when it comes to oil are hardly limited to the right. While right-wing discourses, especially those that adopt a "might is right" approach to the defense of the homeland, are both more prominent and less troubled by the ambiguous and unpalatable outcome of petrorealism—support for antidemocratic oligarchs being the least of these—there are both liberal and left responses to and employment of the discourse of strategic realism. In *Blood and Oil*, for instance, Michael Klare explores the consequences of the US dependency on foreign oil, drawing attention to the huge sums of money that are spent annually to keep access to oil open. He argues that "ultimately, the cost of oil will be measured in blood: the blood of American soldiers who die in combat, and the blood of many other casualties of oil-related violence."[15] For Klare, the proposed solution is for Americans to "adopt a new attitude toward petroleum—a conscious decision to place basic values and the good of the country ahead of immediate personal convenience."[16] The reality of continued growth in energy use in circumstances in which oil is disappearing isn't at issue. Rather, what is proposed is a potentially less violent and more stable way of managing the geopolitical realities created by struggles over access to energy, including vast reductions by Americans in their individual energy usage. The nation remains the central actor, and the misfit between supply and demand for oil is one that needs to be seriously considered so that existing differentials of national power are maintained into the indefinite future. As for the larger consequences of oil usage for the environment or for humanity as a whole? Strategic realism recognizes only that oil is essential to capital and capital is essential for the status quo to remain in place in the future. The disaster in this discourse is figured as the mismanagement or misrecognition of geopolitical strategy, such that a commodity essential to state power is no longer available in the abundance necessary for economic growth.

On the left, meanwhile, there continues to be an abiding fascination with the dynamics of capitalist geopolitics, not, it seems, to plot weaknesses and to imagine something beyond it, but because of the inherent interest in the ceaseless rearrangement of the deck chairs on a capitalist ship that seems in little real danger of sinking.

Techno-utopianism

Strategic realism sees the disaster of oil as a problem primarily for the way in which nations preserve or enhance their geopolitical status. A founding assumption is that the political future will look more or less like the present: strategy can't be developed around the promise of new sources of energy but emerges out of plans to capture and control (economically, diplomatically, or militarily) existing ones. A second narrative related to the looming disaster of the end of oil looks to science and technology to develop energy alternatives that will mitigate the end of oil. This form of techno-utopianism can be used as an element of strategic realism, but in practice these narratives are kept discursively distinct. For instance, the text of the AEI barely mentions oil, focusing instead on nuclear energy, clean coal, natural gas, and renewable energies; the strategic military intervention in the Persian Gulf lies outside of this narrative of future alternatives. Whereas in strategic realism the future is imagined as a continuation of the present, the AEI announces a belief in the new future, albeit one secured by existing sources of energy:

"It will take time for America to move from a hydrocarbon economy to a hydrogen economy. In the meantime, there are billions of barrels of oil and enormous amounts of natural gas off the Alaskan coast and in the Gulf of Mexico."[17]

What I am calling techno-utopianism is a discourse employed by government officials, environmentalists, and scientists from across the political spectrum. With respect to the end of oil, it proposes two solutions: either scientific advances will enable access to oil resources hitherto too expensive to develop (the Alberta tar sands, deep-sea reserves, etc.) while simultaneously devising solutions for carbon emissions (exhaust scrubbers, carbon sequestering, etc.), or technological innovations will create entirely new forms of energy, such as hydrogen fuel cells for space-age automobiles. As with strategic realism, its ubiquity today makes techno-utopianism a familiar discourse. It can be employed as mere political rhetoric to defer difficult decisions with negative economic impacts to some distant future, as in Canadian Prime Minister Stephen Harper's 2006 announcement concerning "intensity-based" emissions standards: "With technological change, massive reduction in emissions are possible. . . . We have reason to believe that by harnessing technology we can make large-scale reductions in other types of emissions. But this will take time. It will have to be done as part of technological turnover."[18] Somewhat more convincingly, techno-utopianism also underwrites the activities of those working actively in science and technology, who hope through their work to offset the civilizational blunder of hitching a complex global economy to a non-renewable dirty fuel source fast evaporating from the Earth.

The utopia I have in mind here is the "bad utopia" of future dreamscapes and fanciful political confections—"utopia" not quite just as an insulting slur against one's enemies, but rather as a projection of an alternative future that is, in fact, anything but a "conception of systematic otherness."[19] In "The Politics of Utopia," Fredric Jameson speaks of "one of the most durable oppositions in utopian projection"—that between city and country. He asks: "Did your fantasies revolve around a return to the countryside and the rural commune, or were they on the other hand incorrigibly urban, unwilling and unable to do without the excitement of the great metropolis, with its crowds and its multiple offerings, from sexuality and consumer goods to culture?"[20] Techno-utopian discourses of future alternatives to oil magically resolve this opposition: since the future is undeniably urban, great metropolises are envisioned as leafy green oases, filled with mid-twenty-first-century *flaneurs* and cyclists who move between buildings crowned with solar sails.[21] All of our worst fears about the chaos that will ensue when oil runs out are resolved through scientific innovations that are in perfect synchrony with the operations of the capitalist economy: problem solved, without the need for radical ruptures or alterations in political and social life.

An excellent example of such techno-utopianism can be found in a 2006 special issue of *Scientific American*, "Energy's Future: Beyond Carbon." The issue's subtitle announces its politics directly: "How to Power the Economy and Still Fight Global Warming." The issue presents technological strategies for carbon reduction, new transportation fuels, efficient building design, clean options for coal, possibilities for nuclear power, and so on.[22] The long-term impact of existing energy use—primarily oil—on the environment is the focus here; each article provides a potential solution based on current scientific research and technological innovation. The articles all begin in much the same way, noting first the deleterious environmental effects of existing social and cultural practices, especially those in the developed world, followed by the failures at the level of politics to mobilize and enforce necessary changes to environmental laws and standards. In his introduction to the special issue, Gary Stix writes: "The slim hope for keeping atmospheric carbon below 500 ppm hinges on aggressive programs of energy efficiency instituted by national governments."[23] But since such programs don't seem to be on the horizon, scientific innovation rushes into the gap vacated by public policy. In the coming disaster of oil, technology absorbs and mediates all the risks that might normally unfold at the level of the political. A profusion of developments from the astonishing to the relatively banal—new refrigerators use one-quarter of the energy of their 1974 counterparts, LCD computer screens 60 percent less than CRT monitors—will bring about not only a cleaner environment but a soft landing for oil capital. If the various timescale charts and projections for reductions in oil usage are less than comforting, we are reminded of the following: "Deeply ingrained in the patterns of technological evolution is the substitution of cleverness for energy."[24] The natural temporal flow of scientific discovery will resolve the energy and environmental problems we have produced for ourselves.

The notion of technological evolution lies at the heart not only of techno-utopian solutions to the disaster of oil but of modern imaginings of science more generally. Technology is figured as just around the corner, as always just on the verge of arriving. Innovation can be hurried along (through increased grants, for instance), but only slightly: technological solutions arrive just in time and never fail to come. In a perversion of Marx's comments in the preface to *A Contribution to a Critique of Political Economy*, it would appear that mankind produces only such disasters as technology can solve; the disaster arises only when the conditions in which to repair it are already in the process of formation.[25] This is, as we see above in Harper's comment, certainly part of the political dream of techno-utopianism. It is equally part of the scientists' self-imaginings as well: "The vast potential of this new industry underscores the importance of researching, developing, and demonstrating hydrogen technologies now, so they will be ready when we need them."[26] At the core of the notion that technological developments are on the horizon to address even such massive, global problems as the end of oil lies a further temporal imagining. If technological developments are thought to be poised to imminently bring about a change from oil capital to (in this case) hydrogen capital, it is because technological developments in the past have always appeared in the nick of time to help push modernity along. But where? And how? History offers no models whatsoever: the fantasy of past coincidence between technological discovery and historic necessity simply reinforces the bad utopianism of hope in technological solutions to the looming end of oil.

Apocalyptic Environmentalism

In his editorial in *Scientific American*, Gary Stix writes:

> Sustained marshalling of cross-border engineering and political resources over the course of a century or more to check the rise of

carbon emissions makes a moon mission or a Manhattan Project appear comparatively straightforward. . . . Maybe a miraculous new energy technology will simultaneously solve our energy and climate problems during that time, but another scenario is at least as likely: a perceived failure of Kyoto or international bickering over climate questions could foster the burning of abundant coal for electricity and synthetic fuels for transportation, both without meaningful checks on carbon emissions.[27]

Narratives of the end of oil that focus on this other scenario are best described as eco-apocalypse discourses. If strategic realism is largely a discourse of the right, its left complement is located largely in eco-apocalypse discourse. These take the disaster of oil capitalism head on: the deep political and economic investments in oil are assessed, the dire social-political environmental consequences of inaction on oil are laid out, and because it becomes obvious that avoiding these results would require changing everything, apocalyptic narratives and statistics are trotted out. Strategic realism and techno-utopianism remain committed to capitalism and treat the future as one in which change has to occur (new geopolitical realignments, innovations in energy use) if change at other levels is to be deferred (fundamental social and political changes). Eco-apocalypse sees the future more grimly: unlike the other two discourses, it understands that social and political change is fundamental to genuinely addressing the disaster of the end of oil—a disaster that it relates to the environment before economics. However, since such change is not on the horizon or is difficult to imagine, it sees the future as Bosch-like—a hell on Earth, obscured by a choking carbon dioxide smog.

The volume *The Final Energy Crisis*, edited by Andrew McKillop and Sheila Newman, is but one of many books and articles in this genre.[28] With great care, clarity, and attention to the scientific evidence about fossil fuel depletion and environmental impacts, the volume lays out the case for getting serious about the looming disaster. The statistics pile up to paint an alarming picture. Fertilizers are impossible to produce without fossil fuels; in their absence, the Earth's carrying capacity for human life will necessarily fall by 50 to 60 percent; the growth in car ownership in India and China to Western levels, even with conservative estimates as to distance traveled, would require 10 billion barrels of oil each year, *"three times the total oil imports of all EU countries in 2002,* nearly three times the maximum possible production capacity of Saudi Arabia"; the postoil population carrying capacity of France is estimated as 20 to 25 million and in Australia less than 1.5 million; and so on.[29] Everything in the volume points to the coming disaster that is the only possible outcome of oil capitalism.

At issue is not the veracity of such claims, which are here always presented relatively conservatively, but what such information is intended to accomplish. All three of the discourses delineated in this essay make claims on the social, inviting it to participate in the framing of a response to the end of the energy source around which we structure social reality—and social hope, and social fantasy. Unlike the other two, the discourse of eco-apocalypse understands itself as a *pedagogic* one, a genre of disaster designed to modify behavior and transform the social. The McKillop and Newman book is exemplary in this regard, combining serious scientific articles (replete with charts and even equations) with Spinozian *scholie*-like passages by McKillop that narrate the coming end.[30] Even while recognizing the potential traumas for human communities and for capital, strategic realism and techno-utopianism operate within existing understandings of the way the world operates. Eco-apocalyptic discourse makes it clear that disaster cannot be avoided without fundamental changes to human social life. With hope for a new way of doing things, the conditions for avoiding disaster are put forward: "A simpler, non-affluent way of life"; "more communal, cooperative and participatory practices"; "new values" ("a much more collective, less individualistic social philosophy and outlook"); and, of course, "an almost totally new economic system. There is no chance whatsoever of making these changes while we retain the present consumer-capitalist economic system."[31]

The difficult question of how such a complete transformation of social life is to be brought about remains open. At best, the reality of a coming future disaster is imagined as being enough on its own to produce the shift in everything from values to economic systems that would be necessary to counter it. There is a form of "bad" utopianism at work here too. Although a new social system is outlined in utopian fashion (down to what kinds of houses should be on a single street and the kinds of animals that we might find in our suburbs[32]), the subject roaming through this landscape is none other than a liberal one, motivated by pleasure, convenience, and comfort. Despite the demands and claims for changing individual behavior and social reality, at the heart of eco-apocalyptic discourses is a recognition that even if its coming can be established, nothing can be done to stop the disaster from coming. Indeed, there is a sense in which disaster is all but welcome: the end of oil might well be a case of capitalism digging its own grave, since without oil, current configurations of capital are impossible.

The Left and Oil Capitalism: Retort and Disaster

National futures, technological futures, and apocalyptic ones. We can, as a form of critical activity, point to the limits of such discourses—to the revival, for instance, of nations and nationalism in strategic realism, or to the shaky temporality of techno-utopianism, or to the political limits of eco-apocalyptic discourses. However valuable such criticisms might be, the issue of what kind of response would frame this disaster in a manner that would create alternatives to oil capitalism still needs to be addressed more forcefully.

The very possibility of a disaster on the scale of the end of oil seems, on its own, not to be able to generate the kind of social transformation one might expect would be needed in order to head off a crisis that would be felt at every level—including that of capital accumulation and reproduction. Jacob Lund Fisker notes:

> The increase in human wealth and well-being during the past few centuries is often attributed to such things as state initiatives, governmental systems and economic policies, but the real and underlying cause has been a massive increase in energy consumption. . . . Discovering and extracting fossil fuels requires little effort when resources are abundant, before their depletion. It is this cheap "surplus energy" that has enabled classical industrial, urban and economic development.[33]

With the end of "surplus energy" thus comes the collapse of surplus profit—or so one would think. It may be that the disaster of oil is already prefigured in the temporal shift of the capitalist economy that goes by the name of neoliberalism. The ferocious return of primitive accumulation, now directed not only toward the last remaining vestiges of the public sector (such as universities and hospitals) but also inward into subjectivity, announces, too, a temporal recalibration of capital away from the future to the present. There is no longer any wait for surplus or any attention to the reproduction of capital for the future; instead, as if the future of capital is in doubt, profit taking has to occur as close to immediately as possible, whatever the long-term consequences.

Something like this view of contemporary capital informs the collective Retort's arguments in *Afflicted Powers: Capital and Spectacle in a New Age of War*. The book as a whole is intended to be a rallying cry for a new left vanguardism that emerges out of the book's framing of the post-9/11 political landscape. This landscape is one structured by a "military neoliberalism"[34] that is described as "no more than primitive accumulation in disguise",[35] this neoliberalism in turn operates largely unopposed due to the dynamics of the "spectacle," which, as in so many appropriations of Guy Debord's concept, appears as a social situation defined by advertising and consumer images—that is, ideology through image form as well as image content.[36] Oil figures as a prominent part of Retort's account of the contemporary political situation. While the authors are struck by the bluntness of the slogan "No blood for oil," as it appears to directly name the reasons for the use of US military force in Iraq, they take pains to argue that placing oil at the causal center of the war is misleading. "Oil's powers," they write, "are drawn from a quite specific force field having a capitalist core that must periodically reconstitute the conditions of its own profitability."[37] The idea that the US invasion was prompted by a kind of petro-Malthusianism—of the kind, it must be said, that informs discourses of strategic realism—is premised on a false assumption about the market for oil. "The history of twentieth-century oil is *not* the history of shortfall and inflation, but of the constant menace—for the industry and the oil states—of excess capacity and falling prices, of surplus and glut."[38]

For Retort, the argument against oil as the cause of the war in Iraq allows the authors to draw out the broader motive driving the use of the military today, which is to support "'extra-economic' restructuring of the conditions necessary for expanded profitability—paving the way, in short, for new rounds of American-led dispossession and capital accumulation."[39] This in turn permits them to consider our contemporary political options against capitalism. Rhetorically, the book makes use of anxiety about the war in Iraq to draw its readers into broader consideration of the dynamics of neoliberal globalization and possible responses to it. The final chapter, "Modernity and Terror," is both where Retort comes clean about its political aims and where it runs up against the limits of envisioning an end to oil capital. The argument the authors make in this chapter is a powerful one. For the left, the opponent is nothing less than the "disenchantment of the world"—modernity itself. There are two central processes that they associate with capitalist modernity. The first is a consumerism that functions by seeming to offer a solution to modernity's disenchantment: "It promises to fill the life-world with meanings again, with magical answers to deep wishes, with models of having and being and understanding (undergoing) Time itself."[40] In other words, commodity fetishism, figured here as the lack of social resources that would allow us to recall the "mere instrumentality" of objects "in a world of meanings vastly exceeding those that any *things* can conjure up."[41] The second process is the "process of endless enclosure,"[42] a continuation of the long process by which natural and human resources were taken from the common for the exclusive use of capital. The goal they set for themselves is to set out a "non-nostalgic, non-anathematizing, non-regressive, non-fundamental, non-apocalyptic critique of the modern."[43] They admit: "The Left has a long way to go even to lay the groundwork of such a project . . . but it is still only from the Left that a real opposition to modernity could come."[44]

Despite the grandeur of such a goal, who could disagree with such a project? Or perhaps just as important: how is this modernity any different from the capitalism that the left has been opposed to all along, even if consumerism and the processes of enclosure are both more intensive and more extensive than in previous eras? One thing that is glaringly absent is any consideration of future disaster. Though Retort pushes oil to the sidelines in its attempt to bring those chanting "No blood for oil!" into its larger critique of neoliberalism, when it considers the function of oil in relation to capital it only looks backward at the history of the twentieth century and not toward the horizon of the disaster that oil's absence will create. Recall Fisker: "It is this cheap 'surplus energy' that has enabled classical industrial, urban and economic development."[45] Oil is hardly incidental to capital or to modernity—which is not the same as saying that it is the prime mover of all decision making by nation-states or other actors in the global economy. At the same time, the growing sense of this coming horizon and the necessity of having to respond to it—whether through the machinations of resource strategy or by leaving it to technology to figure things out—cannot be simply left aside in shaping responses to the dark modernity sketched out by Retort. It is telling, for instance, that there is not even an appeal to the discourse of eco-apocalypse—barely anything at all about environmental limits, population carrying capacity, the need to think up to and beyond oil capitalism. Retort proposes a left response—typically and understandably sketchy and open-ended (how could it not be?)—to the violence of military neoliberalism. But as for a left discourse on oil capitalism that would go beyond the pedagogic gestures of eco-apocalyptic discourse, we have yet to find it.

Can such a discourse even exist? Retort suggests that opposition to what it terms "consumer metaphysics" is rooted in a crisis of time. "What is the current all-invasive, portable, minute-by-minute apparatus of mediation," the authors ask, " . . . if not an attempt to expel the banality of the present moment?"[46] The hope, drifting throughout the social, is for "another present—a present with genuine continuities with a retrieved past, and therefore one opening onto some non-empty, non-fantastical vision of the future."[47] Such futures—futures that are in a very real sense "post"-modernity—are in the process of being created planetwide and in those very spaces where enclosure is violently taking place and consumer metaphysics is at its weakest. As Mary Louise Pratt points out, "Where identities cannot be organized around salaried work, consumption, or personal projects like upward mobility, life has to be lived, organized and understood by other means. People generate ways of life, values, knowledges and wisdoms, pleasures, meanings, hopes, forms of transcendence relatively independent of the ideologies of the market."[48] These narratives of meaning can take many forms, from classic left narratives to wild new religions like the one Pratt discusses, Alfa y Omega, whose two primary symbols are the lamb of god and a flying saucer.[49] Whether such futures are "non-empty" or "non-fantastical" is open to question, even if one was careful to resist measuring them by the standard of whether they figure disaster, much less imagine a way of addressing it.

Whither capital? Will the end of oil capital bring an end to capital as such (and thus, potentially, in its wake, bring new political possibilities)? The expectation that haunts the future is not the end of capital, but that, despite everything, oil capital will not end until every last drop of oil (or atom of fossil fuels) is burned and released into the atmosphere. Fredric Jameson's often-repeated suggestion that "it seems to be easier for us today to imagine the thoroughgoing deterioration of the earth and of nature than the breakdown of late capitalism" points to a limit in how, to date, we have framed the coming future and its disasters.[50] It is not that we can't name or describe, anticipate or chart the end of oil and the consequences for nature and humanity. It is rather that because these discourses are unable to mobilize or produce any response to a disaster we know is a direct result of the law of capitalism—limitless accumulation—it is easy to see that nature will end before capital. As Jan Oosthoek and Barry Gills write, "What is most urgently needed . . . is not short-term technological fixes but a different paradigm of political economy. This new political economy must take our impact on the planet's environment fully and realistically into account."[51] Easy enough to say, but much, much harder to produce when what is called for is full-scale retraction against the flow of a social whose every element moves toward accumulation and expansion.

NOTES

1. "A new nuclear power plant would have to open every few days to replace the world's fossil fuel use in a century, and the problems of renewable, low-density, hard-to-store, distant renewable energy sources will take a lot of time and money to overcome on the scale needed." Julie Jowett, "Fossilised Myths: Fresh Thinking on 'Dirty' Coal," *Guardian Weekly*, Mar. 17–23, 2006, 5.

2. Marx, "Critique of the Gotha Program," 525.
3. The literature addressing the end of oil is now extensive. For some representative studies, see Deffeyes, *Beyond Oil*; Goodstein, *Out of Gas*; Heinberg, *The Party's Over*; Roberts, *The End of Oil*. For a contrary view, see the study by Wood, Long, and Morehouse, "Long-Term World Oil Supply Scenarios."
4. These figures come from Jaccard, *Sustainable Fossil Fuels*. For other studies that take the view that oil remains relatively abundant, see Wood, Long, and Morehouse, "Long-Term World Oil Supply Scenarios," and the especially influential "2000 U.S. Geological Survey World Petroleum Assessment." Wood, Long, and Morehouse place peak oil production as late as 2047, while the USGS estimates that there remain as much as 2.3 trillion barrels of usable oil on Earth (including reserves, reserve growth, and undiscovered reserves).
5. Retort, "Blood for Oil?" in *Afflicted Powers*, 38–77.
6. Lenin, *Imperialism: The Highest Stage of Capitalism*, 11–26.
7. Yergin, "Ensuring Energy Security."
8. Harvey, *The New Imperialism*, 1–25 and 183–212; and Smith, *The Endgame of Globalization*, 177–209.
9. See "Energy Security for the 21st Century," https://georgewbush-whitehouse.archives.gov/infocus/energy/ (accessed August 17, 2018).
10. Alexei Barrionuevo, "Bush Says Lower Oil Prices Won't Blunt New-Fuel Push," *New York Times*, Oct. 13, 2006, https://www.nytimes.com/2006/10/13/business/13fuel.html (accessed August 17, 2018).
11. "China's share of the world oil market is about 8 percent, but its share of total growth in demand since 2000 has been 30 percent. World oil demand has grown by 7 million barrels per day since 2000, of this growth, 2 million barrels each day have gone to China. India's oil consumption is currently less than 40 percent of China's, but because India has now embarked on what the economist Vijay Kelkar calls the 'growth turnpike,' its demand for oil will accelerate." Yergin, "Ensuring Energy Security," 72.
12. As just one example, consider the recent report of a "blue ribbon" task force on US energy policy prepared for the Council on Foreign Relations, which focuses on solutions to US dependency on foreign oil. "The task force suggests that energy security has not been a central focus of U.S. foreign policy, though it noted the widespread perception that the invasion of Iraq and other interventions in the Middle East have been driven by the desire to control the region's oil supplies." Shawn McCarthy, "Report Slams U.S. Domestic Energy Policy," *Globe and Mail*, Oct. 13, 2006.
13. Smith, *Endgame of Globalization*, 188.
14. Yergin, "Ensuring Energy Security."
15. Klare, *Blood and Oil*, 183.
16. Klare, *Blood and Oil*, 182.
17. See the summary of the Advanced Energy Initiative on the White House's Energy Security Web page, https://georgewbush-whitehouse.archives.gov/ceq/advanced-energy.html (accessed August 17, 2018).
18. Prime Minister Stephen Harper, quoted in Bill Curry and Mark Hume, "PM Plans 'Intensity' Alternative to Kyoto," *Globe and Mail*, Oct. 11, 2006.
19. Jameson, "The Politics of Utopia," 36.
20. Jameson, "The Politics of Utopia," 48.

21. See Brown's illustration in *Scientific American* 295, no. 3 (2006): 51.
22. "Energy's Future: Beyond Carbon," *Scientific American* 295, no. 3 (2006): 46–114.
23. Stix, "A Climate Repair Manual," 49.
24. Socolow and Pacala, "A Plan to Keep Carbon in Check," 52.
25. The original reads, "Mankind only sets itself such tasks as it can solve; since, looking at the matter more closely, it will always be found that the task itself arises only when the material conditions for its solution already exist or are at least in the process of formation." *The Marx-Engels Reader*, 5.
26. Ogden, "High Hopes for Hydrogen," *Scientific American* 295, no. 3 (2006): 101.
27. Stix, "A Climate Repair Manual," 49.
28. McKillop and Newman, eds., *The Final Energy Crisis*. See also Darley, *High Noon for Natural Gas*; Deffeyes, *Hubbert's Peak*; Goodstein, *Out of Gas*; Heinberg, *The Party's Over*; Kunstler, *The Long Emergency*; and Roberts, *The End of Oil*, among others.
29. McKillop and Newman, eds., *The Final Energy Crisis*, figures from 7, 232, and 265–73, respectively.
30. See, for instance, the following chapters from McKillop and Newman, eds., *The Final Energy Crisis*: "Apocalypse 2035," 186–90; "The Chinese Car Bomb," 228–32; "The Last Oil Wars," 259–64; and "Musing Along," 289–94.
31. Trainer, "The Simpler Way," in McKillop and Newman, eds., *The Final Energy Crisis*, 280, 283, 286–87, and 284, respectively.
32. Trainer, "The Simpler Way," in McKillop and Newman, eds., *The Final Energy Crisis*, 281.
33. Fisker, "The Laws of Energy," in McKillop and Newman, eds., *The Final Energy Crisis*, 74.
34. Retort, *Afflicted Powers*, 72.
35. Retort, *Afflicted Powers*, 75.
36. For a more philosophically thorough treatment of the concept of the "spectacle," see Jappe, *Guy Debord*, 5–30.
37. Retort, *Afflicted Powers*, 54.
38. Retort, *Afflicted Powers*, 59.
39. Retort, *Afflicted Powers*, 72.
40. Retort, *Afflicted Powers*, 178.
41. Retort, *Afflicted Powers*, 179.
42. Retort, *Afflicted Powers*, 193.
43. Retort, *Afflicted Powers*, 185.
44. Retort, *Afflicted Powers*, 185.
45. Fisker, "The Laws of Energy," 74.
46. Retort, *Afflicted Powers*, 183.
47. Retort, *Afflicted Powers*, 183.
48. Pratt, "Planetary Longings: Sitting in the Light of the Great Solar TV," 212.
49. Pratt, "Planetary Longings," 207.
50. Jameson, *The Seeds of Time*, xii.
51. Oosthoek and Gills, "Humanity at the Crossroads: The Globalization of Environmental Crisis," 285.

CHAPTER 5

Neoliberals Dressed in Black, or, the Traffic in Creativity (2010)

In communist society, where nobody has one exclusive sphere of activity but each can become accomplished in any branch he wishes, society regulates the general production and thus makes it possible for me to do one thing today and another tomorrow, to hunt in the morning, fish in the afternoon, rear cattle in the evening, criticize after dinner, just as I have a mind, without ever becoming hunter, fisherman, herdsman or critic.

—Karl Marx, *The German Ideology*

It is now impossible to tell an espresso-sipping artist from a cappuccino-gulping bunker.

—David Brooks, *Bobos in Paradise*

With the publication of *The Rise of the Creative Class* (*RCC*) in 2002, Richard Florida became almost instantly an influential figure across a range of fields and disciplines. An academic by training, over the past decade Florida has advanced ideas that have shaped discussions of current affairs and the decisions made by businesses and governments. Although he did not invent the term "Creative Class," his thorough analysis and description of the characteristics and function of what he sees as this newly hegemonic socio-economic group guaranteed that he would be identified as its progenitor and primary spokesperson. Florida has remained a staunch defender and advocate of the Creative Class and its related concepts (creative cities and creative economies) over a series of follow-up books that answer criticisms and provide further nuance to the central ideas developed in *RCC*[1]; for him, the financial crash only further confirms the need to place creativity at the center of how we imagine the economy.[2] Nevertheless, it is the first book that remains the most significant, in terms of the articulation of the concepts and ideas he continues to advance, the attention and criticism it has generated, and its lasting impact on the language in which contemporary economic and urban planning decisions are framed.

In Canada, Florida's ideas have generated more praise than criticism, more acceptance than dismissal. His appointment in 2007 at the University of Toronto's Rotman School of Management as Professor of Business and Creativity, and as Academic Director of the newly established Prosperity Institute, was celebrated by local and national media alike. Here was an example of just the kind of Creative Class migration that Florida himself wrote about, with the bonus being that his move from Washington, DC to Toronto seemed to confirm the latter's growing importance as a creative city. Even before his physical arrival in Canada, the discourse of creative cities had been taken up fervently by city governments anxious to find an urban planning narrative to match the challenges and expectations of a neoliberal age. If organizations such as the Creative City Network of Canada (CCNC) or the series of Creative Places + Spaces conferences organized by the non-profit group Artscape are any indication, the idea that creativity is essential to economic growth has been swallowed whole by urban governments across Canada—in big cities such as Vancouver and Montreal, but also in smaller places from Moncton, New Brunswick, to Moose Jaw, Saskatchewan[3]; the Canada Excellence Research Chairs program to bring highly coveted scientific and medical researchers to Canada suggests that the federal government also believes in the economic impact of innovation and creativity. For artists and arts and cultural groups, this attention to the material conditions of creativity might not seem to be a problem. In an effort to create urban environs attractive to members of the Creative Class, local, regional, and national governments have created new programs to support and encourage culture. Instead of being a drain on economies, around the world the arts and culture sector is now seen as a potential financial boon: a segment of the economy in which it is necessary to invest given its overall fiscal impact.[4]

Is there anything wrong with this interest in the economic spinoffs of creativity? Even if only strategically—focusing on the outcome as opposed to the concepts, arguments, and theories employed by Florida and others championing creativity today—doesn't this development represent a productive and positive situation for the arts and culture in Canada (and everywhere else)? If the language of creative cities and the creative class generates more money for museums, increases in grants for artists, expansion of government sponsorship of festivals, support for humanities research on campuses, and so on, what could possibly be wrong with it? I want to argue that the redefinition of culture as an economic resource and as one creative practice among many making up the twenty-first-century economy is a problem. It is not a gain for arts and culture; as recent (2009) cuts to arts and culture funding in both Alberta and British Columbia suggest, the arts continue to be seen as (unfortunately) among the least essential elements of public and social life. But beyond such facts, the expansion of discourses of creativity into the economy at large represents a loss in how we understand the politics of culture—a shift from a practice with a certain degree of autonomy (however questionable, however problematic at a theoretical level) to one without any. In what follows, I offer a detailed analysis of Florida's *RCC* to show what work his discourse of creativity does in relation to the arts and culture. There have now been numerous criticisms made of Florida's ideas, primarily by urban geographers and economists who question his claims about the precise character of Creative Class and the spaces they inhabit. What has not been addressed directly is the very idea of "creativity" on which it all hinges—a concept that has been increasingly called upon to do important conceptual and political work on both right and left.

As is to be expected from a contemporary popular nonfiction text addressing social issues (indeed, it is fundamental to the genre), the core promise and attraction of Florida's *RCC* is its presentation of a new social phenomenon that its author has uncovered; the significance of this phenomenon is figured as being essential to an understanding of the nature of contemporary society, as well as its coming future. The rhetorical form of the book is that of the explorer's tale—the breathless recounting of the discovery of a paradigm shift that reorganizes our very sense of the operations of the social world. Though few others may have grasped it, Florida aims to convince us that the Creative Class is the one primarily responsible for the bulk of economic development today and that its influence on and importance for the economy will only grow in the coming decades. The emphasis on a specific class in relation to its economic function is significant. While Florida presumes to offer a wide-ranging analysis of contemporary society—he positions himself as heir to the work of sociologists such as William H. Whyte, C. Wright Mills, and Jane Jacobs—at its heart this is a labor management book. In the context of a variety of social changes and developments, especially the coming-to-be of the technological society, *RCC* analyzes the characteristics of the Creative Class—their motivations, pleasures, habits, tendencies, goals, likes, and dislikes—in order to give companies the conceptual tools with which they might better capture the fruits of their employees' creativity. It is also a book designed to offer economic advice to city councils and urban planners.[5] Florida makes it abundantly clear that it is not enough to change the work environment of the Creative Class to improve the bottom line. The energies of the Creative Class can be harnessed only in urban environments in which this class finds it appealing to live. The book offers guidelines for the character and nature of the cultural amenities and urban characteristics that provide the preconditions for the creativity so essential to economies today.

It is this aspect of the book that has received most of the critical and media attention directed Florida's way. The long fourth section, "Community," offers an account of what constitutes Creative Centers[6] and an overview of the various statistical procedures he and his colleagues have used to map out the new urban geography of class in the United States.[7] Florida explores the logic of the growing gap between those cities with large numbers of the Creative Class and those without; this division correlates directly with the current financial status of the cities in question. The main question that organizes his examination of urban economics is why members of the Creative Class choose to live in some cities more than in others. A clarification of which characteristics make those cities high on the creative cities index—San Francisco, Austin, Seattle, Boston—so attractive to the Creative Class is intended to assist those at the bottom of the list—Memphis, Norfolk, Buffalo, Louisville—to develop programs and policies to improve their economies.

One can understand why criticism might be directed here. First, local media seized on Florida's book to either trumpet the standing of their cities or dispute it. (Are Buffalo or Memphis really such terrible places to live? Can such places really make themselves attractive to software engineers and financiers?) Second, challenges were made to the veracity and utility of the new indices Florida used to generate his rankings. In addition to indices such as innovation (measured by patents per capita) and high-tech ranking (the Milken Institute's Tech Pole Index), he also made use of two even more controversial measurements: the Gay index and the Bohemian index. It is the politics many felt to be hidden in these measures that produced controversy. For Florida, the Gay index—the number of gay people in a city or a region—indicates a region's tolerance, while the Bohemian index—"the number of writers, designers, musicians, actors and directors, painters and sculptors, photographers and dancers"[8]—identifies the cultural amenities in a region—less such things, it should be noted, as symphonies and concert halls (for which there are other indices) and more the cutting-edge, indie vibe of a place.[9] What do these factors have to do with urban economies? Florida claims that "artists, musicians, gay people and the members of the Creative Class in general prefer places that are open and diverse."[10] He identifies a high correlation between these various indexes, the numbers of Creative Class in an area, and economic success. For reasons that will become clear momentarily, members of the Creative Class are thought to value lifestyle, social interaction, diversity, authenticity, and identity. Florida proposes the theory that "regional economic growth is powered by creative people, who prefer places that are diverse, tolerant and open to ideas."[11] Cities didn't like being deemed uncool (Memphis, Detroit) or intolerant (St Louis and poor Memphis again); further, the breakdown of creative cities as opposed to uncreative ones along party lines—with creative cities tending to be blue (Democrat) and uncreative ones red (Republican)—made many on the US right suspicious of the real intentions of Florida's study.

The claims and arguments made in the latter part of Florida's book about the relationship between cities and creativity have generated criticism; the earlier, more substantive part far less so. The first three sections—"The Creative Age," "Work," and "Life and Leisure"—offer a detailed examination of the character of the Creative Class. It is here, in other words, that he identifies what makes this class meaningfully a class at all. As the preferences for tolerance, diversity, and openness to ideas already named above might suggest, this is not a class in any objective sense of the term, whether understood in the terms of classical economics (the division of the social world into quartiles or quintiles based on income) or in the Marxist sense of those who sell their labor as opposed to those who purchase it. The Creative Class is first and foremost treated as an economic class. It is an economic class that is brought together, however, not just by the fact that its members occupy certain professions but because they adopt a common *style de vie*, an outlook on life that cuts across and ties together the different registers of work, leisure, self-actualization, and social goods. If one had to capture this mode of being in a word, it is in the adjective that Florida gives to this class: creative. One might then expect a clear definition—even an attempt at one—given the very looseness and indefiniteness of the social meaning of the term, which can at times act as little more than an empty approbative: to label something creative is to offer approval or praise. Startlingly, none is given. Nevertheless, there is a core significance and function for "creative" (adjective) and "creativity" (noun) that emerges in Florida's book. To understand the work that the concept of creativity does for his understanding of the social—and indeed, the work it does more generally today, outside of Florida's book as much as within it— one has to consider the significance of the multiple identifications and associations he proposes for the term throughout the book.

Even though many of Florida's descriptions of creativity appear to operate in the same register (that is, they point to the same noun, the same thing, even if they do so with slight variations), looking at the claims and assumptions made in each case is essential. There are (at least) seven forms or modes of creativity identified in *RCC*:

(a) Creativity is an innate characteristic of the human mind or brain. "The creative impulse—the attribute that distinguishes us as humans from other species."[12] It is an attribute that distinguishes the human as such, although it is also described as "a capacity inherent to varying degrees in virtually all people."[13]

(b) It is a cultural or social characteristic and/or good. Just as with individuals, societies can be more or less creative or can be organized to be conducive to creativity or to limit or prohibit it. The text stands as a warning to the United States to be careful about losing its creative edge to countries such as the United Kingdom, Germany, and the Netherlands, which are doing a better job of being creative.

(c) Creativity is the subversion or breaking of rules: "It disrupts existing patterns of thought and life. It can feel subversive and unsettling even to the creator. One famous definition of creativity is 'the process of destroying one's gestalt in favor of a better one.'"[14, 15] Creativity as subversion is especially important in Florida's re-narration of the social drama of the 1960s and its central place in the constitution of the ethos of Silicon Valley[16] and in technological industries more generally.

(d) It constitutes the key element of certain kinds of work, which cuts across the spectrum of previous definitions and distinctions of labor, that is, white collar, blue collar, executive class, working class. There can be white-collar creative jobs just as there can be blue-collar ones, which is why for Florida it is better to speak of a Creative Class instead of depending on these older, Fordist categories. Creative jobs are challenging and involve problem solving. There is an innate pleasure to this kind of work—it wouldn't even be work except for the fact that you are paid (bonus!). Creative people are attracted to their jobs because of "intrinsic rewards . . . tied to the very creative content of their work."[17] Such work allows one to exercise the innate impulse identified in (a).

(e) Creativity is used as a stand-in term for acts that produce the "new": new ideas, new concepts, or new products. In other words, novelty is creativity (and vice versa).

(f) Creativity is strongly linked to technology. One measure of creativity is the number of patents issued per capita; another is the amount of spending on research and development. Florida identifies Nokia cell phones and the film series *The Lord of the Rings* as "creative products." Although he identifies other fields of endeavor and other products as "creative," there is no doubt that he sees the field of contemporary high technology as a place where it is especially in force.

(g) Finally, creativity is repeatedly identified as a characteristic of work in the arts—work done by those whose activities are named by the Bohemian index. This is an element of the arts that has now expanded to cover other forms of human endeavor as a result of social change, technological development, or simply insight into the productive process: in hindsight, many forms of work were always already creative: "Writing a book, producing a work of art or developing new software requires long periods of concentration."[18] When creativity is described by Florida, the arts are always in the pole position: "[Prosperity] requires increasing investments in the multidimensional and varied forms of creativity—arts, music, culture, design and related fields—because all are linked and flourish together."[19]

At times these varied appeals to creativity stand alone; more often, creativity is described and discussed by linking two or more of these different ideas of creativity together. The chain of associations through which Florida runs these works something like this:

technology (f) is creative because it is full of people who are allowed to be subversive (c) and so create new things (e), all as a result of a new social setting (b) that enables companies to (d) create working conditions to permit this to happen.

Working in a high-tech company and being able to be creative in this fashion is the best of all possible worlds but is nonetheless at heart: (a) an expression of an innate human impulse which the economic world has hitherto squashed underfoot.

Unsurprisingly, the circulation of these multiple ideas of creativity generates an increasing number of tautologies and inconsistencies as the book progresses. "Creativity . . . is an essential part of everyone's humanity that needs to be cultivated,"[20] and yet we are also repeatedly told that there are "creative people" (and so presumably less creative ones, too) and a distinct class whose creativity must therefore be the function of something other than simply being human.

The seventh definition of creativity (g) is without question the most important one in Florida's view. His most substantive definition of the Creative Class identifies its key characteristic to be "that its members engage in work whose function is to 'create meaningful new forms.'"[21] A broad definition, to be sure; the nature of these forms and their function is clarified in his elaboration of the kinds of work that constitute the Creative Class. Drawing on categories from the Occupational Employment Survey of the US Bureau of Labor Statistics, he divides the Creative Class into two component elements: the Super-Creative Core and the Creative Class more generally. The first group includes workers across a wide field of employment categories:

> scientists and engineers, university professors, poets and novelists, artists, entertainers, actors, designers and architects, as well as the thought leadership of modern society: nonfiction writers, editors, cultural figures, think-tank researchers, analysts and other opinion-makers. Whether they are software programmers or engineers, architects or filmmakers, they fully engage in the creative process. I define the highest order of creative work as producing new forms or designs that are readily transferable and widely useful—such as designing a product that can be widely made, sold and used.[22]

The Super-Creative Core is paid to engage in the production of new forms that are transferable and useful. By contrast, while the rest of the Creative Class might at times produce new forms, it is "not part of the basic job description. What they are required to do regularly is think on their own."[23] The second group is just as broad and includes knowledge-intensive workers such as legal and health professionals, financial services workers, lawyers, and those who work in the high-tech industry. Should any of these workers have the opportunity to engage in the creation of new forms in their jobs—everything from new products to new job opportunities—they have the chance to move up to the Super-Creative level, "producing transferable, widely usable new forms" as the main purpose of their professions.[24]

Like commodities, such as oil or coal, or the work of laborers in tax-free zones or *maquiladoras*, for Florida creativity is an economic good. Indeed, it is not just one good among many. As he states directly in the preface to the paperback edition and repeats throughout the book: "Human creativity is the ultimate economic resource."[25] Many might imagine creativity to be a quality or characteristic with intrinsic value—a value that isn't established by markets or through its utility or transferability. Florida sees things differently. For him, "creativity has come to be valued—and systems have evolved to encourage and harness it—because new technologies, new industries, new wealth and all other good economic things flow from it."[26] The contribution made by *The Rise of the Creative Class* is thus twofold. First, Florida plays the role of a lobbyist on behalf of creativity to government, business, and the general public, working tirelessly to get these sectors to recognize the importance of creativity to the economy. And second, in his role as a social scientist, he develops numerous theoretical and empirical schemes to understand better the creative-economic systems that have up until now evolved on their own. His aim is to help encourage and harness creativity, so that with the knowledge provided by social science these systems can operate even better, which will equally benefit nation-states and the lives of those workers whose creativity is currently being wasted in jobs that fall outside of the Creative Class.

A utopian vision, is it not? Who could be against more creativity in the world? And the outcomes that creativity seems to produce: more diverse and tolerant societies, better jobs, and wealth for all.[27] Florida's view of the significance of the Creative Class for our collective future in unambiguous: "We have evolved economic and social systems that tap human creativity and make use of it as never before. This in turn creates an unparalleled opportunity to raise our living standards, build a more humane and sustainable economy, and make our lives more complete."[28] This opportunity has not yet been taken up. Luckily, for this extraordinary future to be realized all that is needed is a completion of the "transformation to a society that taps and rewards our full creative potential."[29] A proud member of the thought leadership of "our" society, Florida is prepared to help light the path and to make a fortune (through his consultancy firm) along the way.

Despite his enthusiasm for the project of rendering the world safer for creativity, Florida's view of a social and economic system nearing perfection functions only to the degree that it fails to address or account for a number of issues that—given his subject matter and the concepts he employs—he cannot leave by the wayside. We can get a sense of these gaps and elisions by looking at the few moments in which he raises concerns or questions about the picture he paints. In a 400-page book that sometimes seems intent on addressing almost everything (Jimi Hendrix and the rise of agriculture, Thomas Frank and the Frankfurt School, Silicon Valley and Florida's own childhood skill at building wooden cars), there are only three moments of doubt or hesitation about the views for which he argues. These are worth citing in full:

[The creative economy] is not a panacea for the myriad social and economic ills that confront modern society. It will not somehow magically alleviate poverty, eliminate unemployment, overcome the business cycle and lead to greater happiness and harmony for all. In some respects, left unchecked and without appropriate forms of human intervention, this creativity-based system may well make some of our problems worse.[30]

My statistical research identifies a troubling negative statistical correlation between concentrations of high-tech firms and the percentage of the non-white population—a finding that is particularly disturbing in light of our findings on other dimensions of diversity. It appears that the Creative Economy does little to ameliorate the traditional divide between the white and nonwhite segments of the population. It may even make things worse.[31]

Creativity is not an unmitigated good but a human capacity that can be applied toward many different ends. The scientific and technical creativity of the last century gave us wonderful new inventions, but also terrible new weapons. Massive, centralized experiments in new forms of economic and social life led to fiascos like the Soviet Union, while here in the United States, free-market creativity has turned out a great deal that is trivial, vulgar and wasteful.[32]

There is no comment offered following the first two quotations; they come at the end of sections, after which Florida's cheerleading enthusiasm resumes unabated. The third warning about the potential dangers of creativity comes in the book's conclusion, in which he directs his energies toward convincing the US public and their governments to recognize and support the Creative Class. There is a meek defense offered concerning the potential for creativity to be put toward totalitarian uses or result in the detritus of consumer culture. Put simply, since creativity is now at the core of the economy and since it is only an increase in resources that will enable the potential to do "good in the world,"[33] creativity remains essential, no matter that its results include everything from the atomic bomb to the doodads lining the shelves of dollar stores around the world.

What emerges in these three passages is what is almost entirely absent in the rest of the book: the political. Why a creativity-based system might make our problems worse is never specified; it also comes as somewhat of a surprise, given the tone and triumphalism of the book, to learn that it is not a panacea. One realizes in reading these passages that little or no mention has been made of poverty, unemployment, or the business cycle—or race and ethnicity, for that matter. Yet these are all crucial factors in shaping the experience of work and one's degree of economic participation. There are other ways to make sense of Florida's list of cities and their existing levels of Creative Class workers, which correlate precisely with poverty, unemployment, race, lack of access to education (required for Creative Class jobs), lack of mobility, and lack of opportunity.[34] These are deeply political issues, not mere externalities or afterthoughts to the system he describes. When ethnicity or immigration is discussed, it is framed by Florida as context or backdrop in an urban setting—urban coloring, in other words, much the same as a good alternative music scene: part of the necessary makeup of a city that allows white Creative Class members to feel good about themselves and the place they live. One of the reasons the political is missing—beyond, that is, that its inclusion would spoil the elegance of Florida's system and its apparently strong correlations between job type and so-called tolerance and diversity—is announced in his response to the problem introduced in the last passage above. When it comes right down to it, the logic of the economy trumps everything, even the possibility of the terrible new weapons that some members of the Creative Class are (without doubt) commissioned to design.

At one level, it would not be going too far to see the absence of the political as the absence of the world in general: the contingencies and challenges that shape economic decisions, civic policies, and urban planning are nowhere to be found. This is one of the reasons, perhaps, that there seems to be a fundamental confusion in Florida's work between cause and effect in imagining how urban spaces operate: nowhere does one have the production of creative city spaces that then attract creative workers (away from other creative cities, one can only imagine), but more the reverse is true, with certain kinds of cities emerging out of historically contingent processes of industry and labor. But it is perhaps more productive to focus on a smaller element of the book that nevertheless captures some of its wider absences; this attention will also bring us back around to the function of creativity in relation to culture. The limits of Florida's construction of the Creative Class and its future promise can be seen in the fact that in a book whose fundamental theme is labor, a real discussion of work is entirely absent. The Creative Class engages in the creation of meaningful new forms. It does so, however, as work, as an activity within corporations and institutions familiar to all of us (the ones that capture Florida's interest are high-tech giants such as Dell, Microsoft, and Apple). Work has a number of social and economic functions. One of the most important of these, the reason why a corporation or institution might hire a member of the Creative Class, is to generate a product or offer a service (transferable and useful, whether material or immaterial). This process is not carried out for the good of humanity but to generate profit. As any fourth grader knows, profit can only be realized if the amount one pays the creative staff (and the rest of the workers) is less than the income that can be generated by means of the product. This sense of work—as part of a system of profit, work as something necessary for life—never appears in *The Rise of the Creative Class*. Instead, it is essential for Florida to make the point that members of the Creative Class aren't motivated by money and that the Super-Creative Core makes even less than their Creative peers.[35] In surveys that he cites, IT workers indicate that work challenge, flexibility, and stability all come before base pay as reasons why they choose their jobs, with many other values (vacations, opinions being valued, etc.) standing only a few percentage points behind.[36] For Florida, this interest in factors other than salary is viewed as a defining element of the Creative Class. Their desire for flexible and open forms of work, which allows them to avoid wearing a tie, to come late to work, or, better yet, to continue to work wherever and whenever (at home, on the subway, while shopping, while reading messages on their mobile devices, etc.) is seen as a sign not just of a new mode of labor freedom but a form of social freedom more generally.

Numerous social and cultural critics have drawn attention to the ways in which this apparent new-found freedom in fact covers up an expansion of the work day from nine to five to every aspect of one's life.[37] For those with a more systemic understanding of the economy and the changes it has undergone over the past two decades in particular, this liberalization of the work environment can be understood as little more than a new mode of labor management, whose overall aim remains that of generating as much profit as possible for companies and shareholders. If workers see their jobs as sites of self-definition, challenge, and freedom instead of the opposite, so much the better for the bottom line! The training of bodies willing to work at any time of the day—and to do so not due to exterior compulsion but because of some imagined, self-defining innate drive—is an easy way of increasing productivity without having to increase pay. Many critics and social commentators have expressed deep worries about the ways in which work in the new economy has come to entirely consume life; Florida expresses no such anxieties and even argues that such worries are overstated and beside the point.[38]

At its core, what is expressed in Florida's book is a fantasy of labor under capitalism: the possibility within capitalism of work without exploitation, of work as equivalent to play. What might give those of us who study the arts and culture pause here is how closely this vision approximates that of the ideal social function and purpose of culture—if in reverse. The aim of the historical avant-garde was to reject the deadened rationality of capitalist society through the creation of "a new life praxis from a basis in art."[39] Florida's characterization of the

Creative Class suggests that this new life has in fact been achieved. The passage to the utopia of a new life praxis was supposed to occur via the transformation of life and work by art, such that art as a separate, autonomous sphere of life was no longer necessary. The division of art and life that first made the autonomous activity called art what it is would be undone through the critical activity of art itself. In Florida's vision of our creative present, work tends toward art by means of changes to the character of labor, partly as a result of technological developments and partly due to what can only be described as new-found enlightenment about the way in which the workplace should be configured. Equating the Creative Class with the activity of the avant-garde might seem far-fetched. It is, however, the fundamental way in which Florida envisions the social function of the Creative Class: as having collapsed different spheres of life together in such a way that what is now finally expressed socially is that innate element of creativity that makes us human and distinguishes us from the beasts. He writes, "we are impatient with the strict separations that previously demarcated work, home and leisure."[40] Luckily, we live at a time when these separations have become undone: "The rise of the Creative Economy is drawing the spheres of innovation (technological creativity), business (economic creativity) and culture (artistic and cultural creativity) into one another, in more intimate and more powerful combinations than ever."[41] And again: "Highbrow and lowbrow, alternative and mainstream, work and play, CEO and hipster are all morphing together today."[42]

This is, of course, more wish fulfillment than actually realized utopia. Florida imagines capitalism to have achieved what the avant-garde had wished to bring about as a means of undoing capitalism. How can this be? What enables and sustains the fantasy of capitalism as an avant-garde—capitalism as having gone beyond itself in the way art once imagined it could—is the concept of creativity itself. The history of the concept of creativity and the changes it has undergone over the centuries is enormously complicated. Suffice it to say that in terms of its recent history, creativity is most commonly associated with the act of generation in the fine arts. The idea that the production of a painting or musical score involves generation *ex nihilo*—the emergence of the new out of nothingness—sprang in part from the individualization of the artistic endeavor at the beginning of the nineteenth century and in part from the break with strictly determined formal categories within which artistic activities were supposed to be carried out. Artists are the model for the creative individual; they are also a model of a kind of labor done for intrinsic purposes and outside of the formal institutions of work.

Over the course of the twentieth century, creativity has come to be associated with all manner of activities: scientific discovery, mathematics, economics, business activity, and so on—anything thought to involve the production of newness of any sort. Despite its residual Romantic humanism, one effect of the expansion of uses of the term is to have rendered creativity into a synonym for originality or innovation. In Florida's use of the term, creativity becomes an act with even less specificity, being understood at times as little more than "problem-solving" of a kind that takes place all the time in work and daily life. Yet it is also essential that in virtually every one of its invocations in his work creativity retain its link to the arts and to (the imagined) freedoms and autonomy connected with such work. This is reinforced by the equivalences Florida repeatedly draws between the work of artists and engineers, musicians, and computer scientists. There would be something critical missing in Florida's account if he was to champion the work not of the Creative Class but of "knowledge workers," the "postindustrial class," the "professional-managerial class," "symbolic analysts," or even "cognitive laborers." "Creative" obscures the work function of this class, transforming it into something much grander and more ideal than just a label for a new category of work in late capitalism. The genius of making use of "creative" and "creativity" in the way that Florida does is to render the world into something comprised—if not today, then just over the horizon—of nothing but artistic activity carried out through different forms of labor (not with paint, but XML; not by videos intended for the artist-run space, but for clients on the internet) and with different ends in mind. The distinctions between engineer, computer scientist, and lawyer thus become something akin to those between painter, sculptor, and filmmaker—variants of the same fundamental creative impulse.

The group that most fascinates Florida are workers in the high-tech industry in places such as Silicon Valley and Austin, Texas. His understanding of the nature of technological innovation, the creation of wondrous new hardware and software, shapes his sense of what constitutes creativity. Although he does not say as much, if artistic work stands as the model of what constitutes creative labor in general, new technology is the mechanism by which it is imagined that creativity can form the life activity of more and more people: innovation can eliminate tedious work, leaving only challenging work behind (the dream of a world without work, returned a half century later in a new form). There is, however, another level at which artistic labor and that of technological industries in which Florida is so interested can be seen as connected. At one point in *RCC*, Florida boasts that the number of people who identify themselves as artists and cultural workers expanded dramatically over the past half century in the United States, from 525,000 in 1950 to 2.5 million in 1999, "an increase of more than 375 percent."[43] He declines, however, to consider how such workers actually make a living, which is understandable since, for him, artists and cultural workers value the opportunity to enact their creative freedom much more than they worry about how they might eat.

Florida may be correct in identifying a connection between artists and workers in the knowledge industries of the "new economy." Where he is mistaken is the precise nature of this relationship. What is being carried over from artist to IT worker through the medium of creativity is the "cultural discount" that has long accompanied artistic labor of all kinds. One of the reasons why most artists aren't able to survive on the fruits of their labor is that it is assumed that they are "willing to accept non-monetary rewards—the gratification of producing art—as compensation for their work, thereby discounting the cash price of their labour."[44] Although IT workers are far better compensated for their work than artists, the adoption of an artistic relation to their work effects a similar labor discount that benefits their employer—even if they believe that the primary benefits are theirs. The characteristics of the postindustrial knowledge worker exemplary of Florida's Creative Class entail being "comfortable in an ever-changing environment that demands creative shifts in communication with different kinds of clients and partners; attitudinally geared toward production that requires long, and often unsocial, hours; and accustomed, in the sundry exercises of their mental labour, to a contingent, rather than a fixed, routine of self-application."[45] We are all artists now, which doesn't mean a life of unfettered freedom and creativity. Rather, it means that if we're lucky the labor of crunching code for long hours can be offset by no longer having to wear a tie to work and by getting to play with your colleagues at the corner foosball table once in a while.

There is much more that one could criticize about Florida's vision of our collective futures. There is the fact, for instance, that despite the link he wishes to make between art and the Creative Class, in the end artists and musicians don't really get to play with the big boys of the IT world. The Bohemian index confirms that they are simply the humus out of which the creativity of technological types grows: just like ethnic diversity, they give a place its color and maybe provide an occasional evening's entertainment. The limited vision Florida has of creativity—the almost complete crowding out, say, of any sense of the intrinsic value, or political or social function, of certain kinds of human activity—is indicated by his use of patents as a means to measure it and the unembarrassed description of creativity as pure utility, transferability, and economic functionality. Florida imagines the gradual expansion of the Creative Class so that it would one day encompass everyone. Who would be left to pull the espressos and cappuccinos so beloved by professors and bankers is unclear. What is clear, however, is that even amidst all the creativity in which the Creative Class and Florida himself engages, there is one "new" thing ruled out from the beginning: an entirely new economic system, one in which work would have a very different social character than liberal capitalism, even at its most utopic, might be able to provide.

Some concluding words are necessary to bring this back around to how we think about art and culture in Canada today. To be clear: Florida's views on creativity are less idiosyncratic than symptomatic. One finds these ideas circulating widely in the culture at large. It is endemic in the language of business and economics. As Paul Krugman writes (to take but one example), "in the 1990s the old idea that wealth is the product of virtue, or at least of creativity, made a comeback."[46] The redefinition of business as art via the concept of creativity might not seem to be an especially worrisome problem for the study of culture.

Creativity was never really a feature of older conceptual vocabularies of cultural study (from Winckelmann to Kant to Lessing) and it is certainly not important in more recent ones. My criticisms of Florida might have pertinence to the ongoing ideological redefinition of work and even of social experience and expectation under neoliberalism, yet only be of minor relevance to the practice of contemporary literary criticism, which has never needed creativity, even if creativity has been tied to the activity of art, literature, and culture in the quotidian vocabulary of the social.

But to this I want to sound two warnings. The first has to do precisely with the significance of these shifts in our social understanding of art and culture. In Florida's worldview, what was once dangerous or revolutionary about art has been fully domesticated. The freedom of the artist with respect to some aspects of the organization of their work has, as I argued above, become a model for work in general. As a result, it is thus only the social or political content and not its social form (as work) that might be threatening or dangerous. But once art becomes universalized through the spread of the discourse of creativity, this political challenge, too, is diluted. It is in the adventurous radicality of the artists exhibited in independent galleries and contemporary art museums that Florida locates the kindred spirits of creative workers in other parts of the economy; established museums that display the classics of Western art don't interest him or the Creative Class in the least. If everyone is participating in the same narrative of social development through creativity—artists and IT workers, professors and bankers—what remains of art is to furnish the capitalist economy with ideas indirectly, through the spark or flash of a new concept that might emerge when a software designer is standing in front of a canvas denouncing technological capitalism. The reign of creativity thus poses challenges for the way in which theory and criticism operate today, even if creativity as such may not be a concept with particular theoretical salience within literary or cultural criticism.

But the challenge or threat goes beyond this. I suggested at the outset that creativity was a concept that was being used by both left and right. This essay has focused on the right's use of creativity, identifying Florida as the chief theorist and champion of an idea of creativity that transforms capitalism from a machine of exploitation into something that enables people to fully employ their innate capacities. But what about the left? There, too, the principle of creativity has come to form an important part of how the present social context is conceptualized. Especially in the work of writers associated with Italian autonomist thought, from Paolo Virno to Michael Hardt and Antonio Negri, the current hegemony of post-Fordist, cognitive or affective labor is seen as making evident what was always already true about work but which has become structurally impossible to ignore today. Social prosperity is dependent on language, communication, knowledge, and creativity—that is, on the "general intellect" that Marx describes in a passage in the *Grundrisse* that has become a key part of contemporary left political philosophy. Although it might seem surprising to say so, the difference between right and left, between Florida and Virno, is not in their analysis of the structure of contemporary capitalism and the social and political developments that have accompanied it so much as the lessons that each draws from it. The post-Fordist labor utopia imagined by Florida is for left thinkers anything but the realization of a world without work; instead, it constitutes a new form of exploitation and perhaps an even more dangerous one given the ideological power of accounts of contemporary work such as that of Florida's. Creativity does flourish in contemporary capitalism, but insofar as it is put to use to generate profit, the potential political implications of this new situation are defused, at least temporarily. For the left, the increasing dependence of contemporary societies on forms of creative labor constitutes a political and imaginative opening—recognition (at long last) that capitalism needs labor far more than labor needs capitalism and that the sovereignty of the state can be replaced by a new society founded on the general intellect.

Yet despite the different lesson left and right draw from the social and political implications of post-Fordist work, they share a surprisingly common view of what constitutes creativity and its links to art, culture, and the aesthetic. In recent social and political thought, creativity seems to have become nothing short of the defining element of human being: we are no longer *homo faber* but *homo genero*. As in the case of Florida, creativity on the left finds its referent in an idealized vision of artistic labor and a skewed view of the character of classical aesthetics and is also imagined as what needs to be enabled and set loose in order for there to be genuine social freedom. In a recent interview, Virno points to the troubling integration of aesthetics into production but in so doing affirms a view of aesthetics that is reminiscent of Florida's own claims about the place of creativity in human nature.[47] While admitting at the outset of this long interview that his knowledge of modern art "is actually very limited," Virno is fearless in extending several of his key concepts to discussions of art and aesthetics, such as "virtuosity," one of many names for the innate productive capacity of human beings, which in the work of Antonio Negri goes by the name "constituent power."[48] If left discourses are attuned to the blind spots that exist in Florida's celebration of the conditions of work under contemporary capital, they nevertheless enact the same rhetorical and conceptual gesture of transforming human activity (or at least its potential) as such into art—and an idea of art taken not from sociology but from fantasies about its ideal relationship to something called creativity.

The effect once again is to render mute the critical capacities and political function of art and culture, even as it becomes coterminous with human life activity as such. Contemporary left theoretical discourse might not result in increased funding for the arts, but its temptations for cultural theory can be just as great—and just as problematic. It places art at the center of politics but only by doing away with the significance of art as art. Contemporary art and cultural production have a social specificity that plays an essential role in their political function. They don't need to think of themselves as creative or as the exemplar of creative acts. Indeed, it would seem that the farther they stay away from the intellectual and political traffic in creativity, the greater suspicion with which they treat this mobile and uncritically accepted discourse, the more likely they are able to continue to challenge the limits of our ways of thinking, seeing, being, and believing.

NOTES

1. See *The Flight of the Creative Class*, in which Florida examines the global competition of states and cities to attract members of this class; *Cities and the Creative Class*, which constitutes an elaboration of his description of the communities creative workers are attracted to and in which they flourish; and *Who's Your City?*,

which puts his analysis to use in the form of a city guide for members of the creative class.

2. See Florida's *The Great Reset* and his article "How the Crash Will Reshape America."

3. Although the CCNC predates Florida's books, its growth and expansion since becoming a not-for-profit organization in 2002 has been enabled by the spread of the idea that city spending on culture supports economic development. The CCNC acts as advocate of and clearinghouse for ideas linking culture and economic development. For example, the January 2010 *Creative City News* reports on the investment of $5 million by the City of Woodstock in the creation of a new art gallery; the December 2009 newsletter includes stories on urban investments in culture in places such as Barrie and Collingwood, Ontario, and Halifax, Nova Scotia.

4. Governments across the world have in recent years produced planning strategies for their cultural sector in relation to its economic impact or have developed new departments of government to manage the economics of culture. To give a few examples: Winnipeg is concluding its year as Cultural Capital of Canada with the production of an arts and culture strategy document, "Ticket to the Future: The Economic Impact of the Arts and Creative Industries in Winnipeg." In the United Kingdom, the Creative and Cultural Skills unit of the national government announced £1.3 million to create two hundred culture jobs for young people claiming unemployment benefits, including positions "such as theatre technician, costume and wardrobe assistant, community arts officer and business administrator."

   The action is just as great on the international level. Numerous international conferences focus on culture and economics, such as the annual Culturelink Conference (the third meeting of which was held in Zagreb, Croatia, in 2009) and the World Summit on the Arts (the fourth meeting held in Johannesburg in 2009). The recently released report of the Commonwealth Group on Culture and Development, a body established in 2009, links the achievement of development goals with the support of culture. And UNESCO's November 2009 World Report, "Investing in Cultural Diversity and Intercultural Dialogue," warns governments against cutting funding to culture during the current financial crisis, not just because it will impact on the issues contained in the report's title but because such fiscal cost saving will have a deep impact on any possible financial recovery.

5. Less so for state or national governments: just as for thinkers such as Saskia Sassen, for Florida, the city is the primary political and economic unit of the contemporary era. See, for example, his review of Thomas Friedman's *The World is Flat*, "The World is Spiky."

6. Florida, *Cities and the Creative Class*, 218.

7. The Canadian edition of *Who's Your City?* extends this analysis to Canada, if in a more limited way.

8. Florida, *Who's Your City?*, 260.

9. "This milieu provides the underlying eco-system or habitat in which the multidimensional forms of creativity take root and flourish. By supporting lifestyle and cultural institutions like a cutting-edge music scene or vibrant artistic community, for instance, it helps to attract and stimulate those who create in business and technology." Florida, *The Rise of the Creative Class*, 55.

10. Florida, *The Rise of the Creative Class*, 250.

11. Florida, *The Rise of the Creative Class*, 249.
12. Florida, *The Rise of the Creative Class*, 4.
13. Florida, *The Rise of the Creative Class*, 31.
14. Florida, *The Rise of the Creative Class*, 32.
15. This quotation is unattributed by Florida, as are a number of others in the book. The likely source for the quotation is Max Wertheimer, one of the founders of Gestalt theory.
16. Florida, *The Rise of the Creative Class*, 190–211, especially 202–10.
17. Florida, *The Rise of the Creative Class*, 87.
18. Florida, *The Rise of the Creative Class*, 14.
19. Florida, *The Rise of the Creative Class*, 320.
20. Florida, *The Rise of the Creative Class*, 317.
21. Florida, *The Rise of the Creative Class*, 68. The quotation that Florida includes here is unattributed.
22. Florida, *The Rise of the Creative Class*, 69.
23. Florida, *The Rise of the Creative Class*, 69.
24. Florida, *The Rise of the Creative Class*, 69.
25. Florida, *The Rise of the Creative Class*, xiii. The number of times this claim is asserted is too numerous to cite, but take for instance statements such as these at opposite ends of the book: "Today's economy is fundamentally a Creative Economy" and "Creativity is the fundamental source of economic growth." Florida, *The Rise of the Creative Class*, 44 and 317.
26. Florida, *The Rise of the Creative Class*, 21.
27. The critical importance of tolerance to manage the perpetuation of hegemony appears in numerous works in the genre of the popular books on current affairs. See, for example, Chua, *Day of Empire: How Hyperpowers Rise to Global Dominance—and Why They Fall*.
28. Florida, *The Rise of the Creative Class*, xiii.
29. Florida, *The Rise of the Creative Class*, xiii.
30. Florida, *The Rise of the Creative Class*, 23.
31. Florida, *The Rise of the Creative Class*, 262–63.
32. Florida, *The Rise of the Creative Class*, 325.
33. Florida, *The Rise of the Creative Class*, 325.
34. Mobility is presumed to be a central characteristic of the Creative Class. They can go wherever they want, which is why cities have to make certain that they have the appropriate environs to attract them. Yet even in the case of certain members of the Super-Creative Core, this mobility is close to a fiction. For example, academics find it extremely difficult to move; the nature of their work means that they have to participate in specific kinds of institutions (universities and colleges) that aren't found in the same proportion as institutions of private industry and many of which are located in smaller cities and towns. There's a reason why Durham, NC, and State College, PA, rank highly on his rankings of creative cities: it's not because they have a huge number of amenities (art, coffee houses, alternative music, etc.) that exist outside of work but because the nature of the institutions that exist there render large number of PhDs (especially relative to population) immobile.
35. Florida, *The Rise of the Creative Class*, 77.

36. Florida, *The Rise of the Creative Class*, 88–101.
37. See, for instance, Fraser, *White Collar Sweatshop*; Marazzi, *The Violence of Financial Capitalism*; Ross, *Nice Work If You Can Get It*; Schor, *The Overworked American*; and Terranova, "Free Labor."
38. "The no-collar workplace is not being imposed on us from above; we are bringing it on ourselves . . . We do it because we long to work on exciting projects with exciting people. We do it because as creative people, it is a central part of who we are or want to be." Florida, *The Rise of the Creative Class*, 134.
39. Bürger, *Theory of the Avant-garde*, 49.
40. Florida, *The Rise of the Creative Class*, 13.
41. Florida, *The Rise of the Creative Class*, 201.
42. Florida, *The Rise of the Creative Class*, 191.
43. Florida, *The Rise of the Creative Class*, 46.
44. Ross, "The Mental Labour Problem," 6.
45. Ross, "The Mental Labour Problem," 11.
46. Krugman, *The Return of Depression Economics and the Crisis of 2008*, 24.
47. "There is an aesthetic base component in human nature." Paolo Virno, "The Dismeasure of Art," np.
48. Paolo Virno, "The Dismeasure of Art"; Negri, *Insurgencies*.

CHAPTER 6

The Cultural Politics of Oil: On *Lessons of Darkness* and *Black Sea Files* (2010)

Capitalist production has not yet succeeded and never will succeed in mastering these (organic) processes in the same way as it has mastered purely mechanical or inorganic chemical processes. Raw materials such as skins, etc., and other animal products become dearer partly because the insipid law of rent increases the value of these products as civilizations advance. As far as coal and metal (wood) are concerned, they become more difficult as mines are exhausted.

—Karl Marx, *Theories of Surplus Value*

For the past two years I've been receiving news feeds from the *New York Times* and other magazines and newspapers alerting me to articles dealing with oil. In some weeks, the number of articles I would receive was staggering: seven or more in a day, fifty or more in a week, and all from a single newspaper. Many of these articles were (predictably enough) about the rise in gasoline prices (which peaked in the United States at $4.11 per gallon in July 2008) and its impact on airline ticket sales, the driving habits of suburbanites, and the price of anything that needed to be shipped to market—which is to say almost everything. Others dealt with existing and emerging geopolitical challenges connected to the cost of oil and the problem of its increasingly limited supply: the rise of petro-oligarchies (from Russia to Venezuela, from Kazakhstan to my home province of Alberta), the stresses placed on energy supply by the expansion of developing economies (especially China and India), and the impact of oil on global food supplies. Finally, there were articles that dealt in a broad way with the ecological impacts of the ever-increasing use of dirty energy such as crude oil. These articles stressed the need to develop new sources of energy and tried to draw attention to the necessity of reshaping, in a fundamental way, our daily habits and practices in order to save the planet.

Oil was everywhere, connected to everything—and yet there was something missing. Despite all that has been and continues to be written about oil, it still seems to be difficult to capture the fundamental way in which access to petrocarbons structures contemporary social life on a global scale. Oil is not one energy source among others—a bad habit that needs to be overcome through the creation of the energy equivalent of a nicotine patch that would slowly wean people off their 84-million barrel-a-day habit and put us on the path of cleaner living and healthier lungs. Oil is not just energy. Oil is history, a source of cheap energy without which the past century and a half would have been utterly different. And oil is also ontology, the structuring "Real" of our contemporary sociopolitical imaginary, and perhaps for this reason just as inaccessible as any noumenon in the flow of everyday experience from the smoggy blur of sunrise to sundown. When one discusses the end of oil and imagines the main issue to be the possibility of replacement fuels—basically, energy from the sun, in whatever form—one fails to grasp that we are not dealing with an input that can easily take other forms, but with a substance that has given shape to capitalist social reality, perhaps as much as the division of labor or the dance of commodity reification.[1]

The cosmic joke is on us: the last two centuries of capitalist social development has burned through energy resources which are the product of 500 million years of geological time. As M. King Hubbert, of the famous Hubbert's Peak, writes:

> When these fuels are burned, their precious energy, after undergoing a sequence of degradations, finally leaves the earth as spent, long-wavelength, low-temperature radiation. Hence, we deal with an essentially fixed store-house of energy which we are drawing upon at a phenomenal rate . . . The release of this energy is a unidirectional and irreversible process. It can happen only once, and the historical events associated with this release are necessarily without precedent and are intrinsically incapable of repetition.[2]

The arcs of population, gross domestic product, and energy consumption over the past century and a half all swoop upward in perfect harmony when graphed against one another. It is the massive increase in per capita energy consumption that "has enabled classical industrial, urban, and economic development."[3] Too bad that what is a temporary source of energy has been treated as permanent and fundamental to our growth economies, and that, even on the brink of a looming disaster, the end of oil tends to disappear over the horizon as the result of indifference, long-established habits, or the difficulty of imagining that things could really be as bad as all the geologists and ecologists say they are; the decrease in the cost of a gallon of fuel due to the global financial crisis has resulted in the immediate return of older patterns of driving.

Earlier I identified three dominant narratives through which the crisis of the end of oil has been described and comprehended to date: strategic realism, techno-utopianism, and eco-apocalypse.[4] Discourses of strategic realism deal with the problem of oil as being primarily about the ways in which governments secure ongoing access to diminishing supplies of energy. Techno-utopianism recognizes that the continuation of our current global social and political reality requires a high level of energy use, and imagines technological solutions that would substitute new forms of energy for those on which we currently rely. Finally, eco-apocalypse discourses—the main form of oil discourses on the left—focus on the need to fundamentally reshape contemporary social life. These discourses are aware of the absolute dependence of society on petrocarbons and try to generate an alarm loud enough to produce a social awakening regarding our plight. They are apocalyptic in a double sense: first, because they are aware of the real nature of oil—that oil is history and social ontology—and are anxious about the implications of its decline for human populations and the massive fixed infrastructure of cities and transportation systems in which they live; second, because despite the ability of these discourses to name the problem, to describe it in detail and with great complexity, they confront a political and cultural impasse that is seen, finally, as being nearly impossible to overcome. Increasingly sophisticated charts detailing the use of land and sea resources, overviews of the coming future mapped through Peak Oil charts and photo-essays showing the impact of the economics of oil on human communities and nature—there is a sense that none of these will do what one might hope they would, i.e., help produce new political circumstances and a clearer idea of the challenges we collectively face.[5] Which is to say: apocalyptic environmentalism is as traumatized by the failure of social rationality—the Enlightenment and its promised forward march from immaturity to maturity—as it is by the material consequences of current patterns of energy use. Or rather, the real limits such discourses confront concern the politics of representation—of producing social and political change from the narratives about the future that they paint—in the face of the positivism of technological thinking or political imaginaries which see resource usage primarily as a problem of national security and the health of domestic economies.

Are these the only ways to think about oil? Or are there are other narratives that go beyond the stubborn Realpolitik of strategic-realism, the magic of technological thinking, or the guilty pleasures of the coming end times? It would be helpful to have as a fourth discourse a Marxism that is engaged not just with ecology (à la Joel Kovel and others) but with the political, economic, *and* conceptual significance of raw inputs into the shape of capitalism.[6] With the exception of some attention to resource scarcity (e.g., the work of Michael Perelman), the tendency, especially with the rising interest in creative or cognitive labor, has been to affirm Marx's view that nature is subsumed directly into production without mediation.[7]

In the absence of such a fourth discourse, but with the problem of political representation in mind, I want to consider the way in which the politics of oil are addressed in two films—Werner Herzog's *Lessons of Darkness* (1992) and Ursula Biemann's *Black Sea Files* (2005).[8] Both usefully complicate my typology of oil narratives and their sterile politics, and offer some insight into both what is missing and what is all too present in each of them. Strategic realism and techno-utopianism are narratives that insist on maintaining the status quo at any cost; eco-apocalypse understands that oil is history, ontology, and culture, but can't see the way forward given oil's omnipresence just below the skin of society (even as it is disappearing beneath the skin of the Earth). Current ways of thinking about oil either ignore or affirm an antinomy that they know to exist even if they can't name it, as opposed to trying to think their way through it. Might these films about oil, and about the way we represent it or fail to represent it to ourselves, give us some ideas about how to generate other narratives about the future to come—an ecological politics that abandons the comforts of either apocalypse or business as usual?

Given the social and economic importance of oil and its strategic importance since at least World War I, it comes as somewhat of a surprise that there are so few films or other cultural narratives that address it head-on. The oil crisis of 1973 produced many more cultural artifacts than our current encounter with steep oil prices—everything from board games to presidential addresses on the need to severely reduce fuel consumption.[9] The two most common filmic forms in which we encounter oil today are geopolitical thrillers, in which oil takes the place vacated by ideology in the Cold War, and documentary films, which carry out a hoped-for pedagogic function in bringing to light the problems caused by our dependence on oil. In thrillers like *Syriana* (2005) and *The Deal* (2005), oil is central to the plot and yet nevertheless incidental. Narratively arranged with multiple storylines that take place in numerous locations—a now common, overly literal attempt to represent the new reality of globalization—the struggle over oil resolves into a fairly standard storyline about the links between corporate and political power, and the ways in which greed and money deform social life. In this respect, it is no different from the television serial *Dallas* (1978–1991), and equally belated. Of slim importance to the plot, oil in *Dallas* is merely the source of the Ewing family's wealth and the reason that they live in Dallas (though one would have thought Houston to be a more appropriate site for oil moguls); a telenovela *avant la lettre*, money, power, and family strife drive the narrative (Lucy won't go to school! J.R. is sleeping around!), which is otherwise so uninterested in geopolitics as to miss the fact that after 1973 oil wealth springs not from the soils of Texas but from other parts of the world (those smart Ewings wasted a lot of money on state senators who were unlikely to be able to help them with oil claims in Saudi Arabia).

On the other hand, documentaries such as *A Crude Awakening* (2006) or *Fuel* (2008) follow a pattern traced out by other recent docs on a range of topics.[10] The consequences of our dependence on oil are analyzed by experts, alarming statistics are trotted out concerning the years remaining in the life of oil and our levels of CO_2 emissions, and we are implored to do something before it is too late. Oil thus becomes the latest in a long list of social problems that could be resolved except for a lack of political will (which seems to be in shorter supply than even hydrocarbons), which the documentarians hope to kick-start by informing the public. Documentaries on oil tend to follow the script of eco-apocalypse; so, too, do oil thrillers, although they do so through a fascination with the blind self-confidence and violence of strategic realism, which offers irresistible resources for drama and characterization.

Black Sea Files and *Lessons of Darkness* undertake comparatively novel explorations of the politics of oil. In part this is because they attend to oil as a social problem that still needs to be puzzled out rather than as a lesson about the fact that corporations want money or that driving produces CO_2 (neither an especially shocking insight). It is also because each engages in experiments in documentary form in order to make sense of the place of oil in our lives. Herzog's better-known *Lessons* is an example of his ongoing work in a genre that has been called "nonfiction feature." Deliberately pushing against the indexical qualities of documentary cinema, he has repeatedly intervened in the construction of purportedly "real" films by staging scenes or inventing story elements to enhance dramatic narrative (as in *Little Dieter Needs to Fly* [1997]), and has made use of documentary and found footage to construct fictions out of documented realities. *Lessons* is modeled on his earlier *Fata Morgana* (1972), a more poetic, less linear film based on footage shot in and around the Sahara Desert, linked together by the director's voice-over reading of a Mayan creation myth and songs by Leonard Cohen. The images in *Lessons* come from footage taken over a month-long period of oil fields left burning in Kuwait at the conclusion of the first Gulf War in 1991. More than 700 wells were set on fire by Iraqi troops at the conclusion of the war; it required eight months to fully extinguish the fires due both to their scale and intensity and because land mines left around the wells had to be identified and defused before crews could move in. An estimated 6 million barrels of oil per day were burned off by the fires; the total cost of putting them out ran to US $1.5 billion. The twelve chapters of Herzog's film drift over the hellish landscape produced by roaring, jetting flames of oil, white sand turned black in every direction, the machines and bodies of insect-like humans rendered insignificant by the scale of the devastation.

Black Sea Files is also arranged into sections: ten "field notes" (numbered 0 to 9) taken as part of a visual exploration of the development of a British Petroleum pipeline running from Bibi-Heibat in Azerbaijan through Georgia and Turkey, where it ends at an oil terminal on the Mediterranean. Biemann's film is both video and art project: while the ten field notes can be shown successively as a documentary, they have also been exhibited in museums and galleries on separate monitors that can be viewed in any order. In *Black Sea Files*, images appear on a split screen, both halves in constant motion so that it is difficult to follow either; voice-overs and right-to-left streaming text complicate matters even more. As in Herzog's film, its episodic character interrupts but doesn't entirely displace a narrative trajectory; Biemann's visual journey from source to mouth of the pipeline jumps forward and backward in space and time as a way of deferring a linearity that would reduce her investigation into mere exposé, but the overall movement is still from source (0) to mouth (9).

If there are similarities at a formal level, the approaches of the two films to the subject of oil are starkly different. *Lessons* is set up as a science-fiction film—a visual document narrated by a visitor to Earth (a trope reused by Herzog in a different way in 2005's *Wild Blue Yonder*). Over steaming oil-soaked piles of sand, the film opens with a voice-over: "A planet in our solar system. White mountain ranges, clouds, a land shrouded in mist." Cut to a workman in a white moon suit, gesturing at the camera: "The first creature we encountered tried to communicate something to us." What we are presented with is thus the view of the Earth and of humanity as a whole as it appears to alien eyes studying and trying to comprehend what organizes the life activity of the creatures it encounters. A city, strangely empty; a war fought over oil that leaves behind the husks of cars, trucks, and the bones of animals on oil-soaked sand; bombed-out refineries and pipelines so rusted that their color seems to have been leeched by the sun over centuries. Much of the rest of the film is made up of endless aerial shots of oil deserts and oil lakes, oil bubbling and boiling on the surface of the Earth, recovery teams laboring with saurian machines to stop the spray of oil into the air, not to plug it up for good but so that they can suck it out of the ground on their own terms. The concluding voice-over makes Herzog's aims evident: "Two figures approach an oil well and set it ablaze again. . . . Has life without fire become unbearable for them? Others, seized by madness, follow suit. Now they are content, now there is something to extinguish again." Oil is at the center of the activity of these creatures; without it, even given its evident destruction of nature and culture, they would not know what to be or how to live.

Lessons brings our dependence on oil to light quite literally: never has the inky black stuff been made so visible, raining from the sky after a well fire has been put out, draining off of workers' helmets in streaks across their faces, shown in Borgesian fashion as equivalent to the territory of the Earth itself. Oil is also at the center of Biemann's film, but what she wants to make visible is not its physical substance, but the social and human geographies it produces. As in previous projects, such as her investigation of gender politics in the structuring of labor in the maquiladoras in the video *Performing the Border* (1999), *Black Sea Files* sets itself the project of writing "the hidden matrix of space," of mapping the lived material realities and everyday experiences produced out of the abstract language of contracts, company planning maps, and handshakes between politicians. On a map, a pipeline runs straight across territories that look empty, devoid of life; Biemann wants to understand what has been pushed aside in order to make this emptiness real and what new space its previous inhabitants are now forced to occupy. The introductory segment (File 0) makes explicit her project: to explore geography as an ordering system, one now organized around not weapons technology but the power of resources. The BTC pipeline project exhibits the geographic and political power of oil in condensed form, as new petrocapitalist states rush to put in the necessary infrastructure to bring product to market. Even on a map, existing ethnic tensions and political legacies are embodied in the path followed by the pipeline. While it runs in long straight vectors, it also takes sharp turns to circumnavigate Armenia and to skirt Kurdish territory in Turkey.

Each of the files in *Black Sea Files* fills in the story of the new geographies of resources. File 1 consists of images of the primary oil extraction site, while File 2 jumps ahead to Istanbul to show a family of Kazakhs who have been displaced from China for ethnic reasons before doubling back to the shallow oilfields of the Caspian Sea; these apparently distinct movements are somehow related. In File 3, we see the pipeline being laid into the ground by a multinational group of engineers and geologists (mostly Colombians), who are creating not just a pathway for oil but a curious geopolitical space: an 80-meter (87.5-yard) wide, 750-kilometer (466-mile) long strip of land subject to no national laws for 40 years. File 5 visits farmers across whose land the pipeline runs. In the sixth file Biemann interviews two prostitutes who have traveled the routes taken by oil tankers from Azerbaijan to Turkey. Oil riches obviously don't trickle down; the completion of the pipeline will endanger even this precarious form of employment once the truckers disappear. The final three files investigate areas around the pipeline's terminus. In File 7, Biemann learns from a social anthropologist who is advising BP that part of the rationale for the project is to reduce the environmental impact of oil tankers traversing the Black Sea and the Bosphorus. The last two files try to make sense of the political and signifying gap that exists between sites of oil production and oil consumption.

Biemann's closing voice-over brings the project to an indeterminate conclusion: "Throughout my investigation, I was bound to visit the secondary scenes of current affairs, roaming around the lesser debris of history." Secondary, lesser: do the visual and geographic investigations in *Black Sea Files* complete some picture of geopolitics that we would otherwise understand only in parts and so not grasp correctly? Is she offering up (in line with a fairly typical documentary imperative) the small picture obscured by the bigger one of geopolitics—a document of a desire just to render visible the minute, invisible processes by which a resource reshapes geography (and so politics, too)? What exactly are we supposed to do with the materials assembled in these files? Biemann herself isn't sure. The video is meant as neither investigative report nor aesthetic artefact; her struggle with the project's political and artistic commitments are made explicit in File 4:

> [*voice-over*] What does it mean to take the camera to the field, to go to the trenches? How did it get to the point where she stands

at the front next to the journalists at the very moment of the incident? Without press pass or gas mask. What kind of artistic practice does such video footage document? That of an embedded artist immersed in the surge of human confrontation and confusion. How to resist making the ultimate image that will capture the whole drama in one frame? How to resist freezing the moment into a symbol?

Is an image made under dangerous conditions more valuable than material found in libraries and archives? Is better knowledge that which is produced at great risk?

It sounds odd, but it's risky to simply record a pipeline. Oil companies run a severe image regime. During construction, image making is prohibited; later it will be invisible anyway. What is the meaning of this tube in the hidden corporate imaginary of this space? What function does it have in their own secret ordering system of the Caucasus?

To generate images of oil infrastructures is not an aesthetic project, it is an undercover mission. The challenge is to go undetected when probing for hidden, secret, and restricted knowledge. Are these cognitive methods any different from the ones used by geologists, anthropologists, or secret intelligence agents?

They all probe different sorts of sediments and plots that give meaning to the space. What is the sediment I should be probing in my artistic fieldwork? What role do I play in this plot?

"To generate images of oil infrastructures is not an aesthetic project, it is an undercover mission," Biemann says. Herzog's *Lessons* would seem to occupy the other side of this dichotomy. We are meant to be impressed, awed, and seduced by the scale of the oil disaster, its stark colors (red on black, white on black), and the dynamic visual energy of shooting flames and roiling clouds of smoke. The film paints the present as an apocalypse we are not only fated to live in, but in which we apparently find comfort as well: without the burning oil, social life would lose its dynamic force, its animating rationality. The narrative voice of the film (an alien, remember) struggles to make sense of what it sees: can there really be such creatures for whom oil is an object of worship? But Herzog can't resist commentary, and so in parts of the film we get excerpts from Revelation (e.g., 16:18: "And there were voices, and thunders, and lightnings; and there was a great earthquake, such as was not since men were upon the earth, so mighty an earthquake, *and* so great"), which direct us back to the film's epigraph, attributed to Blaise Pascal but actually an invention of Herzog's: "The collapse of the stellar universe will occur like creation—in grandiose splendor." This seems to be the judgment of history and thus, strictly speaking, outside of politics—an expression of a profound gloom about human possibility that confirms the conservative inclination of the alien metaphor, which has become an increasingly common device in recent art (not to mention the remake of *The Day the Earth Stood Still* [1951 and 2008], now also about nature rather than the Cold War). The view from an alien perspective—the making literal of a meta-perspective—is intended to allow structural truths to emerge without the intervention of morality, ethics, or political confusion. But this can often backfire: instead of insight, a radical incommensurability between subject and object opens up, resulting in a disavowal of the human as such, what J. J. Charlesworth, writing on the alien, calls "a pessimistic apprehension of impending disaster; a profound sense of uncertainty and disorientation regarding human society's claim to progressive agency; and a kind of post-historical estrangement from the experience of modernity."[11]

The passages from Revelation aren't the only breaks with the science fiction narrative established at the beginning of *Lessons*. There is the music, for one thing: selections from Grieg, Mahler, Pärt, Prokofiev, Wagner, and others, which amplify the visual scale and lend the film its Romantic gravitas. The chosen scores are ones now cemented into social consciousness through their use in popular cinema to gesture to sublime experiences—the overwhelming encounter with fate, fear, or otherness. In this instance, however, consciousness does not return to itself empowered, but with a sense of the impossibility of overcoming humanity's oil ontology. Then there are the two chapters—Chapter IV "Torture Chamber" and Chapter VI "Childhood"—in which the camera shifts its focus from the oil fields to the victims of the Gulf War, that is to say, to the victims of oil. "Torture Chamber" begins with a tracking shot over what are presumably a collection of torture devices found after the war: knives, pliers, vices, whips, electrical devices, even a toaster. The chapter ends with the mute testimony of a woman who tries to talk to the alien narrator about the horror of watching her sons tortured to death in front of her—but she has lost her ability to communicate. "Childhood" is also about the inability to communicate the experience of a trauma. A young boy whose head was crushed beneath the boot of a soldier says to his mother "Mama, I don't ever want to learn how to talk" and remains permanently silent thereafter. It may not be an "undercover mission" in the way that Biemann imagines *Black Sea Files*, but, like that film, *Lessons* betrays an uncertainty about how to explore its topic and best make sense of all the footage collected in Kuwait. Herzog makes this failure of communication explicit diagetically so that it doesn't appear as an aesthetic limit or failure of his approach, but as the very problem that he hopes to foreground in the film.

On a straightforward reading, Biemann's seems to be the more politically astute film—the one which avoids the grand drama and pretense of metaphysics and speaks about the complex shapes of life lived under the reign of oil. What connects the projects is the joint frustration they exhibit about the object which they want to represent and better comprehend. Both Biemann and Herzog take oil as the name for a complex problem which requires formally innovative methods of exploration if one is to do more than produce an already known object lesson about fuel consumption and the evils of SUVs; it is simultaneously a problem that raises questions about the function of an aesthetics with political aims and intentions. It is as interesting to take note of what is absent from these films so as to assess the ways in which oil is thematized and visualized. There is no state to rail against, and no weak politicians in the pocket of big business to expose: they make a brief appearance in *Black Sea Files*, but only as part of a larger system that exceeds them: the need for oil to flow. The objective here is not the animation of (that fantastically suspect concept of) political will, whether of audiences or of public figures. Even in a video that takes its project to be an investigative one, the results of the investigation are indeterminate: Biemann generates an analytic of Black Sea oil geographies, but we're given no sense of how or even *if* we are supposed to use this knowledge for some form of political intervention. The traumatic futures to come as a result of the end of oil are also nowhere evident. One billion barrels of oil disappeared in flames in Kuwait—which simply means that more secure access to oil is needed (such as the BTC pipeline) in order to avoid the expense of future wars over the resource.

There is one other thing missing: nature. It seems that all discussions and analyses of our use of resources and of oil inevitably bring natural systems into play. It is nature that is seen as bearing the cost of a global social system built on oil. The sites and spaces of oil production generate enormous amounts of pollution; tanker spills—widely underreported—leave behind the bodies of dead animals and zones of ocean and shoreline no longer fit for animal and plant habitation. It is of course the larger, systemic effects that raise the greatest alarm. Oil is a problem for nature because the emissions released from burning it produce changes in the atmosphere whose impact on both natural and human systems are likely to be significant and difficult to undo. Oil narratives are thus often about what humans must do in order to mitigate or limit these effects. Yet neither film seems to have much interest in pursuing this script.

Biemann is interested in social and political geographies as opposed to physical ones, in the way in which social systems are bent to make the flow of this commodity possible. And Herzog? A desert drowned in oil might seem to be a direct comment on the human impact on the environment. But I think this would be to misread the film, especially given Herzog's view of nature. The human relation to nature has emerged as a theme in a number of Herzog's films, the most well-known being *Grizzly Man* (2005). In Les Blank's documentary on the making of *Fitzcarraldo* (1982), Herzog expresses the contrarian view of nature, which has guided his filmmaking up to and including his most recent film, *Encounters at the End of the World* (2007):

Taking a close look at what is around us, there is some sort of harmony. It is the harmony of overwhelming and collective murder. And we in comparison to the articulate vileness and baseness and obscenity of all this jungle, we in comparison to that enormous articulation, we only sound and look like badly pronounced and half-finished sentences out of a stupid suburban novel—a cheap novel—and we have to become humble in front of this overwhelming misery and overwhelming fornication and overwhelming growth and overwhelming lack of order. Even the stars up here in the sky look like a mess. There is no harmony in the universe. We have to get acquainted to this idea that there is no real harmony as we have conceived it.

But when I say this I say this all full of admiration for the jungle. It is not that I hate it, I love it, I love it very much. But I love it against my better judgement.

Lessons treats nature in this way: both with admiration and suspicion, insisting on its disorder, its mess. Oil *is* nature here, not something that should be banished from it as a foreign element. What the spew of oil draws to our attention are the problems that exist in our comprehension of nature that would see the bodies of zooplankton and phytoplankton as something completely other, an alien substance that just happens to lie below the surface of the Earth and that fulfills—what luck!—two key requirements for its use by capital: it can be easily transported and stored, and it generates a significant amount of energy per unit of fuel. In Herzog's imaginary, we love oil, and we love nature, too: both organize our activity and sense of human purpose. For him, it is the desire for harmony and purpose in nature that fuels our love for it; it is a desire that must be guarded against if we want to understand our deep cultural and political imperatives—the how and why of the bodies shuffling across the black sand who can't help but worship the black ink and the heat and light it generates.

Unbehagen in der Natur: a play on the original German title of Freud's *Das Unbehagen in der Kultur* ("The Uneasiness/Discomfort in Culture," translated in English as *Civilization and its Discontents*). This is the title of the final chapter in Slavoj Žižek's *In Defense of Lost Causes*. For Žižek, our contemporary uneasiness about nature is certainly justified; and it is a productive discomfort—as long as we understand it for what it is. He worries that the "ecology of fear"—our worries about everything from the potentially disastrous outcome of biogenetic experiments to anxieties about the exploitation of Earth's resources—"has every chance of developing into the predominant form of ideology of global capitalism, a new opium for the masses replacing declining religion."[12]

Why might this be the case? Like Herzog, Žižek views the treatment of nature within most forms of ecological thinking as fundamentally conservative. Nature is treated as the ultimate form of order that (in the last instance) offers security to human social life; there is also an insistence on the fact that the natural world is complete unto itself and that any change with respect to it "can only be a change for the worse."[13] The position Žižek argues for is that of "ecology without nature," since he feels that "the ultimate obstacle to protecting nature is the very notion of nature we rely on."[14] This ecology without nature would be one that starts from an acceptance of the fact that "'nature' *qua* the domain of balanced reproduction, of organic deployment into which humanity intervenes with its hubris, brutally throwing its circular motion off the rails, is man's fantasy; nature is already in itself 'second nature,' its balance is always secondary, an attempt to bring into existence a 'habit' that would restore some order after catastrophic interruptions."[15] It is in the chaos and groundlessness of "second nature" that any political act that has a hope of radically confronting ecological catastrophe has to take place.

Might we not see these films as attempts at producing "ecologies without nature"? Oil seems like the most basic of substances; worries about what we might do without it seem as easily addressed as the placement of ads for wind and solar power in the pages of magazines, or shifts in public policy (which have been taking place, if not in the United States or Canada, then in Germany and France). If I have treated oil as something stranger, full of metaphysical mystery and subtlety, it is because it is in a very real way absent from social life—despite the fact that it is all around us in physical form in plastics, fuels, fertilizers, and so on. In both the labor theory of value and in the language of economics, the resources we depend on are strictly speaking without value: for capitalism, nature always comes for free. It is only the cost of extracting resources—the cost of ground rents, labor, and materials—that appear on ledger sheets, which is why it has proven so difficult for ecological economists to give a number to nature's contribution to economic processes. Estimates of the economic contribution of nature

range from US $36 trillion annually to infinite; the point of such cal-
culations is to generate changes in social behavior by making the real
"costs" of our actions on the ecological system part of the system of
value.[16] But how do you "price" a finite input that is essential to the
operation of the whole system? Aren't oil and capitalism in a sense one
and the same? These questions are difficult enough to properly pose,
much less to answer. No grand conclusions, no mysteries solved: *Black
Sea Files* and *Lessons of Darkness* draw attention to the desperate need
for contemporary left theory to engage in the difficult work of making
oil and other natural resources a central part of our political imaginings
and strategizing, and of the need to do so without the comforting ease
of dreams of transcendence and salvation.

NOTES

1. For what it's worth, in an effort to highlight our reliance on energy, Jean-Marc
Jancovici has (provocatively) invented a "slave equivalent" measure to point to
how many laboring bodies worth of energy (calculated quite precisely in kilo-
watt-hours) an average person makes use of in her daily activities. The energy
consumption of the average French person is equivalent to each owning 100 slaves
to work for her (cooking, cleaning, generating heat, moving them around, etc.);
the average American would require closer to 200 slaves. See http://www.manicore
.com/anglais/documentation_a/slaves.html (accessed August 17, 2018).
2. Quoted in Marsden, *Stupid to the Last Drop*, 49.
3. Fisker, "The Laws of Energy," 74.
4. See chapter 4 of the present book.
5. For examples of all of these, see Knechtel, ed., *FUEL*.
6. Kovel, *The Enemy of Nature*.
7. "As soon as he has to produce, man possesses the resolve to use a part of the available natural objects directly as means of labour, and, as Hegel correctly said it,
subsumes them under his activity without further process of mediation." Marx,
Grundrisse, 734.
8. *Lessons of Darkness*, directed by Werner Herzog (1992; Troy, MI: Anchor Bay
Entertainment, 2002), DVD; *Black Sea Files: Video Essay in 10 Parts*, directed by
Ursula Biemann (2005; Video Databank).
9. See Borasi and Zardani, *Sorry, Out of Gas*.
10. Other examples include *The End of Suburbia* (2004); *Peak Oil: Imposed by Nature*
(2005); *The Curse of Oil* (2005); *The Power of Community—How Cuba Survived
Peak Oil* (2006); *An Inconvenient Truth* (2006); *Who Killed the Electric Car?* (2006);
Blood and Oil (2008).
11. Charlesworth, "Any Other But Our Selves."
12. Žižek, *In Defense of Lost Causes*, 439.
13. Žižek, *In Defense of Lost Causes*, 441.
14. Žižek, *In Defense of Lost Causes*, 445.

15. Žižek, *In Defense of Lost Causes*, 442.
16. For the oft-cited former figure, see Constanza, et al., "The Value of the World's Ecosystem Services and Natural Capital." For a contrary view, one that emphasizes (among other things) that possible stresses to the natural world are already accounted for by standard measures in capitalist economies, see Lomborg, *The Skeptical Environmentalist*. See also Jacob Stevens's insightful review of Lomborg, "Monetized Ecology."

CHAPTER 7

Crude Aesthetics: The Politics of Oil Documentaries (2012)

How does the problem of oil appear in documentary film? In what follows, I examine the manner in which oil is represented in three "feature" documentaries released over the past five years: Basil Gelpke and Ray McCormack's *A Crude Awakening* (2006), Joe Berlinger's *Crude: The Real Price of Oil* (2009), and Shannon Walsh's H_2Oil (2009).[1] As might be expected, while each has oil at its core, these documentaries differ substantially in both subject matter and form. Berlinger's *Crude* deals with a protracted legal case against the activities of Chevron in Ecuador; Walsh's film examines the ecological and social impact of the Alberta oil sands, specifically their effects on the communities that rely on the water used in conjunction with bitumen processing; and Gelpke and McCormack offer an overview of the politics and economics of oil, together with the environmental damage it causes and the potential crisis of the end of oil. By examining them together, I want to consider the range of ways in which these documentaries frame oil as a problem for their audiences, and what resources they offer as possible solutions to this (historically unprecedented) social and ecological problem. These documentaries both reflect and are a source of the social narratives through which we describe oil to ourselves; it is revealing to see both the limits and the possibilities of the narratives they proffer, which are pieced together out of the fragments of concepts and discourses dating back to the Enlightenment concerning nature, the social, and human collectivity.

As is the case with documentaries on a wide range of social issues, these films about oil understand themselves as important forms of political pedagogy that not only shape audience understanding of the issues in question, but also hope to generate political and ecological responses that otherwise would not occur. This production of an outcome or change in societal imperatives is a long-standing desire of the kind of politically and ethically committed documentary filmmaking that for publics has to a large degree become identified with the function of documentary as such—even if there may be relatively scant evidence of the hoped-for translation of audience awareness of film themes into political action outside the theater.[2] While it nonetheless remains productive to critically assess the political efficacy of documentaries like these—whether by considering the formal or stylistic approaches each makes to its subject matter,[3] examining their capacity to effectively expose "the gap between self-professed norms and behavior,"[4] or by probing the generic politics of such "commodity biographies"[5]—my aim here will be to consider what these documentaries tell us about the social life of oil today. In what follows, I will treat these films as providing examples of narrative and aesthetic choices through which the problem of oil is framed—or can be framed—not only within the films, but within the social more generally; the site of politics I will focus on is not the success or failure of any given documentary to constrain or mobilize a political response, but rather what the discursive, narrative, and aesthetic strategies employed suggest about the dominant ways in which the problem of oil is named and solutions to it proposed. Fredric Jameson famously describes cultural texts or artifacts as "symbolic acts" in which "real social contradictions, insurmountable in their own terms, find a purely formal resolution in the aesthetic realm."[6] It is in this sense that I will offer readings of these three documentaries as aesthetic acts that, in their own specific manner, have "the function of inventing imaginary or formal 'solutions' to unresolvable social contradiction"—unresolvable in perhaps a stronger and more determinate way than the social contradictions to which Jameson referred.[7]

My essay proceeds in three parts. First, by offering readings of these documentaries, I draw out the ways in which each narrates the social life of oil. In her recent discussion of human rights films, Meg McLagan argues that these are developed around the axiom that to expose hidden forces and problems to the light of film is to generate the capacity in publics to address the situations the films uncover.[8] One of the reasons for focusing on these three films in particular is that while they, too, might have this axiom at their core, they proceed with the awareness that the importance of oil to social life is already well known, that publics have yet to adequately respond to its demands and looming crises, and indeed that they may be entirely unable to respond even if they adequately understand the issues. As my analysis of these three films will show, the "solutions" these films offer to the social contradictions generated by oil are made difficult by the fact that the place of this resource in our lives seems to defer the politics one hopes to generate from the production of a documentary about it—and not just the politics directly connected to documentary practice, but to broader ideas that persist about the relationship between belief and action in the operations of social life more generally. In the second part, I draw out some key discursive and conceptual claims made within these documentaries about the unprecedented social problem of oil. Finally, I conclude with an exploration of exactly what kind of "unresolvable social contradiction" oil might be. In "Two Faces of the Apocalypse," Michael Hardt productively explores the antinomies that define and separate the anticapitalist and environmental movements.[9] The insights offered by these films suggest that the problem of oil has the potential to destabilize the aims of both movements. As surprising as it may sound, it is the socially taken-for-granted physical substance of oil—and, of course, the practices that it supports and enables—that has to be placed conceptually and discursively at the heart of both movements if either is to realize its ambitions.

Oil on Film: *A Crude Awakening: The Oil Crash*; *Crude: The Real Price of Oil*; and *H₂Oil*

An increasing number of documentary films address the role, function, and impact of oil in the world today. The three films that I am discussing here attempt to map the social ontology of oil—the how, why, and wherefore of oil in our social, cultural, and political life. *A Crude Awakening* alerts publics about the degree to which contemporary global society is dependent on a natural resource necessarily in short supply. *Crude* and *H₂Oil* each examine the environmental consequences of oil exploration, with a focus on its effects on those indigenous communities who live in proximity to the resource and who thus have to endure both the ecological traumas of ongoing drilling and the sludge and slurry left at past drill sites. What distinguishes H_2Oil from *Crude* is that the former includes brief lessons on peak oil as part of its overall narrative and makes this an element of its case against the Alberta oil sands; *Crude*'s focus, on the other hand, is on the dynamics of law and corporate power as these play out in relation to a commodity at the heart of capitalism's profit logic. In what follows, I probe the "lessons" each provides for thinking about oil by drawing out the (implicit or explicit) ways, both thematically and formally, in which they address the problems this substance generates.

A Crude Awakening: The Oil Crash is divided into ten sections (introduced by intertitles) that provide a narrative of the significance of oil for contemporary global society. It takes the form of a social documentary intended to identify and explain a contemporary problem hidden from view. The secret exposed here is the depth of dependence of contemporary social and economic systems on oil—a non-renewable resource whose era of abundance and easy access is now past, even if this fact seems little acknowledged by the manner in which it continues to be used and exploited.

The film conveys the gravity of our historical moment with respect to oil through three techniques. First, it showcases testimonials about oil from a large number of experts. The range of expertise on which the filmmakers draw is impressive, as is the attention to the politics of each of these talking heads. Two of the most prominent speakers are Matthew Simmons, an energy investment banker, author, and adviser to President George W. Bush; and Roscoe Bartlett, a Republican Congressman from Maryland. The film is careful to include voices from the oil industry, as well as from academics and scientists who deal with the issues the film raises from the vantage point of their own specialties; notable for their absence—with the sole exception of attorney Matthew David Savinar, who until recently ran a website on the politics of peak oil—are those activists or environmentalists (or even Democrats!) whom one might expect to find in a film awakening us to the challenges of peak oil.

A second technique is the communication of information about oil through the use of facts and statistics. These come directly from the mouths of the experts themselves, and they are invariably alarming (e.g., each calorie one eats require ten calories of fossil fuels to produce; by 2010 the planet will have to bring 200 million new barrels of oil on stream *per day* in order to deal with the depletion of existing wells as well as growth in demand; and so on). Finally, the context of peak oil is framed through the formal decisions made with respect to the images and sounds that fill up the space between the talking heads. There are numerous points one could make with respect to the particular use of montage and fast-cutting in many of the sequences in the film. The speed of much of the visual evidence, especially against the backdrop of Philip Glass's minimalist soundtrack, suggests "a life out of balance," as do the many images meant to evoke oil culture: sheiks walking through fancy shopping malls, sludge-filled rivers and oceans, battlefields on which wars have been fought over oil, and the mess of drill sites all over the world. At times, *A Crude Awakening* interlaces these images with older footage of car ads, instructional videos, and clips from celebratory corporate documentaries, all of which appear in hindsight not just as shortsighted but as obscene testaments both to humanity's waste and (in the case of the clips from the instructional videos) to the very different relationship between supposedly objective knowledge systems (i.e., science and documentary film) and oil in the not too distant past.

The ten sections of the film build an effective case against oil. They link oil to geopolitical conflict (Section 4: A Magnet for War), identify its centrality to daily social life (Section 2: We Use it for Everything!), and explore the reasons for concern about the end of oil (Section 6: Peaking Out). What it does not do is offer a solution or resolution to the coming oil crash. The third section of the film looks at three spaces of oil production that have experienced the traumatic passage from oil boom to bust (McCamey, Texas; Maracaibo, Venezuela; and Baku, Azerbaijan). These are micro-case studies intended to provide examples of what might soon happen on a macro-scale. What we see are images of formerly flourishing towns and cities, now semi-abandoned and ugly. The images of Baku's oil fields (which have been captured iconically in the photographs of Edward Burtynsky[10]) are especially haunting: the screen is filled with the remnants of old wooden derricks running up and spilling into the Caspian Sea, fresh oil still staining the ground. If these cases are meant as object lessons, one might expect them to be followed by information as to how it might be possible to manage the down cycle of oil that will soon be experienced on a planetary scale. *A Crude Awakening*, however, seems intent on informing its viewers that there is no way of offsetting a planetary crisis. The penultimate section of the film (Section 9: Technology to the Rescue?) presents possible options—electricity, hydrogen, biomass, nuclear, wind, and so on—only to have technology experts rule each of them out on the basis of inefficiency (e.g., at present it takes three to six gallons of gas to create enough hydrogen to enable us to drive the same distance as one gallon of gas), scale (10,000 nuclear plants would be needed to replace oil), or lack of resources (with that many nuclear plants in existence, uranium reserves would be exhausted in one to two decades). The film lays open the consequences of a civilization based on oil in order to present audiences with some insight into the why and how of the conflicts and pressures of the near future—a future about which there is little of substance that can be done due to the weight of existing infrastructure and the realpolitik of power in contemporary political and economic systems.

There appears to be a deliberate decision in *A Crude Awakening* to avoid directly linking the narrative of peak oil to the impact of petrochemicals on the environment. The question whether or not continued oil use—either at current or at higher levels—will damage the environment is suspended, one suspects, in order to focus on the necessity of oil to current ways of living and being, and to preclude challenges to the film that might emerge from the growing contingent of climate skeptics. By contrast, H_2Oil and *Crude* each explore specific examples of the impact of oil exploration and production on the environment and human communities. What we learn from these cases is not only the manner in which oil damages both ecological and human health, but also the degree to which the interests of elected governments, national legal systems, and multinational corporations are intertwined in ways that make difficult the possibility of addressing some of the specific (as opposed to systemic) impacts of oil. *Crude* examines a landmark legal case against the consequences of the oil exploration and extraction conducted by Texaco (purchased by Chevron in 2001) in Ecuador from 1964 to 1993. There are two main anchoring narratives in the film. The first follows the actions of Ecuadorean lawyer Pablo Fajardo and his American counterpart Steven Donziger over a two-year period (2006–7) as they pursue a suit against Chevron on behalf of thousands of members of the Cofán indigenous community. The second is a single moment in the trial in which plaintiffs, defendants, and the presiding judge in the trial visit the Lago Agrio oil field as part of the evidentiary process. In the first narrative, we witness the political and cultural struggles in which Fajardo and Donziger engage in an effort to generate awareness and legitimacy for their case. In addition to on-the-ground fights within the Ecuadorean legal system, this includes actions at Chevron shareholders' meetings, talks with the New York legal firm that is funding the suit, and engagement with the (then) new left-wing government of Rafael Correa. In January 2007, the public-relations battle they conduct in conjunction with the legal proceedings is accelerated as a result of the commission of a *Vanity Fair* article on Fajardo's fight against Chevron on behalf of the Cofán, which leads to the involvement of pop singer Sting and his wife Trudi Styler, and results in Fajardo being given a CNN Hero's Award in 2007. Even though the plaintiffs build legitimacy and support for their case in the media, legal maneuvering by Chevron means that a case that had at the time of the film's release (2009) already been in process for fourteen years would continue for another ten: the documents collected in the trial's evidence room are so numerous that it is difficult to imagine any judge being able to work through them in a meaningful way even in the decade estimated by the film at its conclusion.

The perspective of the film is clear: Chevron is at fault and is using its immense power as a multinational corporation (US \$204 billion in revenue in 2010) to make a conclusion to the trial impossible. The dirty soil and water, and the numerous health problems of the Cofán (infant deaths, cancers, skin lesions, and more), contrast starkly with the talking-head segments with Chevron scientists and lawyers, whose mobilization of scientific data attesting to the safety of their drill sites cannot but seem little more than corporate lies (indeed, the film points out that Ricardo Reis Veiga, the Chevron lawyer interviewed in the film, was indicted for fraud by the Ecuadorean government). Despite the fact that Donziger is shown to work the system in sometimes ethically questionable ways (he whispers to Trudi Styler to mention Chevron as frequently as possible in her comments on the situation of the Cofán, and the New York firm for which he works stands to make a fortune if the case is successful), his relentless indictment of Chevron's corporate malfeasance mirrors the film's own perspective on the situation both in Ecuador and in the world at large.

However, the second guiding narrative of the film complicates this easy indictment of Chevron's actions. In this section, Fajardo and Chevron's attorney, Adolfo Callejas, move around the Lago Agrio oil field, each making points as to what might constitute physical evidence for use in the trial (contaminated water, oil-soaked soil, and so on). While Callejas uses numerous tactics to shield Chevron from responsibility for the site, all return ultimately to the question of ownership. Callejas argues that while Fajardo and the plaintiffs make numerous claims, they provide no substantial evidence. Chevron disputes the claim that the water is contaminated by oil, or argues that such contamination as does exist introduces no health risk; it insists that it is impossible to link water contamination to oil that they own (as opposed to oil that might have seeped into river water or groundwater from other drill sites or through natural means); and they make numerous legal points in relation to property rights. Property begets responsibility, and so Chevron argues that Petroecuador assumed responsibility for the site when they took it over, that the site was always a Texaco–Petroecuador consortium (such that the latter shares whatever responsibility is assumed for the former); that Texaco no longer exists as a company and so cannot be held responsible; and that the area in which the Cofán live was designated an oil exploration site by the government in the 1960s, and so no people should be living there to begin with. Taken together, these points (and there are others in a similar vein) offer a confusing defense. Rather than building a coherent case, it is as if they are being thrown out in the hope that one or another will stick. After all, if there is no pollution, then does it matter who owns the oil? If it were truly the case that Petroecuador has had responsibility for the site since 1992, why would Chevron be anxious about the level of pollutants in the area?

And does the government not bear ultimate responsibility for the Cofán if it has allowed this indigenous group to live in an area not intended for people? So why mount a defense about "safe" levels of oil in water *or* who owns what, when? From the perspective of the film, such confusing and overlapping arguments constitute further evidence against Chevron. But, from another perspective, the claims made by Callejas and other Chevron employees draw attention to the metanarrative of the film, which is less about oil than about the constitutive, systemic gap between, on the one hand, social responsibility, equality, and justice, and, on the other, the legal and political mechanisms that are in place to address the very real crisis faced by a community that now lives on in the barely concealed sludge of former drill sites.

As its title indicates, H_2Oil is also about what happens when oil finds its way into water as a result of industrial oil extraction. In the main, this film looks at the effects of the Alberta oil sands on the First Nations (Athabasca Dene) community in Fort Chipewyan, a hamlet situated on Lake Athabasca near the terminus of the Athabasca River. The Athabasca runs through the primary site of bitumen extraction and constitutes an important element of the process by which oil is recovered from the near-solid "tar" that makes up the oil sands. Based on the recorded levels of polycyclic aromatic hydrocarbons (PAH) and arsenic in the water of both the river and the lake, the Dene and environmental scientists argue that the Athabasca River is absorbing the chemicals left behind by the extraction process, whether through deliberate action or through errors and accidents in the retention of tailings. The main body of the film moves back and forth between claims and counterclaims about the level of toxins in the Athabasca by the Dene and the Alberta government, and in so doing explores the larger dynamics of corporate and political power in the province as it follows attempts by members of the Fort Chipewyan community to draw attention to the serious environmental and health problems they face.

While it is committed to the exploration of the problems of Fort Chipewyan, H_2Oil makes use of this case to outline the larger political, economic, and ecological entanglements generated by the oil sands. Well-known critics and commentators on Canada's oil policy (and its connection with climate change), such as Tony Clarke, Dr. David

Schindler, and Dr. Gordon Laxer, are given an opportunity to weigh in on the implications of current government decisions (or lack thereof) on greenhouse gas emissions, water and soil contamination, and national resource independence. There are also short cartoon segments included that provide quick instructional overviews of the mechanics of oil sands, the implication of the Security and Prosperity Partnership of North America for Canadian water and oil, and the places to which the end product of the oil sands are pumped (*all* of the oil is currently exported to the United States). If *Crude* emphasized the role of corporations in the narrative of oil and water, the antagonists in H_2Oil are in the main government agencies and ministries, whose representatives argue that they are behaving in a responsible and efficacious manner to address health and ecological concerns. The Ministry of the Environment disputes every one of the facts and figures on cancer rates, oil seepage, and carcinogen levels in the water proffered by scientists critical of their practices. A secondary narrative concerning the problems generated by a drill site for a spring water company based in Hinton, Alberta, amplifies this criticism of government, highlighting how difficult it is even for businesspeople outside the oil industry to bring attention to the overuse and contamination of groundwater as a result of oil exploration and extraction.

Notable for its absence in H_2Oil is the oil industry itself. With few exceptions, its presence is signaled only by the frequent images inserted into (what have become) a form of generic montage about the oil sands: enormous, glowing refineries, made up of systems of pipes, exchangers, and condensers of almost unimaginable complexity; slow aerial pans of the vast extraction sites, framed against the edges of boreal forest now fast vanishing in their wake; and the slow-motion movement of grasshoppers (oil pumping units) conjoined with (in a fashion similar to *Crude Awakening*) sped-up images of consumer modernity—driving, building, shopping. The film is careful to highlight the close connection between industry and government in Alberta. The Office of the Environment is located in the Petroleum Tower in downtown Edmonton, and the assistant deputy minister of the Oil Sands Sustainable Development Secretariat, Heather Kennedy, is identified as a former employee of oil giant Suncor. Nevertheless, in contrast to the intimacy with which H_2Oil engages with the Dene and others (e.g., Fort Chipewyan's medical doctor, John O'Connor), oil corporations are filmed at a distance, figured as inhuman Goliaths in comparison to the all-too-human Davids living in Northern Alberta who are dependent on water that makes them sick.

Taken together, these three films and the critical discourses that they mobilize—multiple in each case, and neither dogmatic nor simplistic—provide insight into how the problem of oil is framed and negotiated, both within documentary and beyond it. These investigations of oil on film generate three insights into the discourses and narratives of the politics and problems of oil. Earlier I argued that there are three broad social narratives through which the futures of oil (and so approaches to its present) have been articulated: strategic realism, techno-utopianism, and eco-apocalypse.[11] These three documentaries interest me in particular because they do not fall easily into any of these categories; nor are they examples of the kind of formally inventive, reflexive documentary on the problem of oil to which I have devoted attention elsewhere.[12] While they share some of the conclusions of these latter documentaries, their commitment to a more expository or observational documentary form places them to one side of my earlier taxonomic scheme—neither abandoned to the realpolitik of struggles over diminishing resources, nor advocating a miraculous technological solution, nor accepting the disastrous fate of the end of oil even while critiquing the manner and extent to which we late moderns use it. Even while they are cautious not to promote "solutions" to oil (even in the case of *Crude* and H_2Oil, films for which redress for the effected indigenous communities might constitute at least a small step forward), they avoid the (sometimes too easy) discourse of eco-apocalypse. In all three films, conclusions are suspended in order to better map the nervous system of oil capitalism.

System Failure, Antinomy, Scale

Though these films open up our understanding of oil across a wide range of conceptual and aesthetic registers, three seem to me especially important: system failure, antinomy, and scale. I will take each of these up in turn.

System failure

All three films make clear that our existing social systems are inoperative. Though it might seem obvious to say it, oil is only a problem because of the larger systems through which it flows. The injustices faced by the Dene and the Cofán cannot and will not be resolved through existing mechanisms of law, property, electoral politics, or knowledge (i.e., science). The struggles waged by both indigenous groups regarding the scientific establishment of levels of pollutants in their water highlight the malleability of knowledge when it bumps into the imperatives of government and business. Systems of property and ownership overwrite questions of corporate or ethical responsibility: one rejoinder by Chevron lawyer Callejas is that it is impossible to identify the oil in the Cofán rivers as belonging to Texaco because "it doesn't have a trademark on it." There is no suggestion in *Crude* that a different legal outcome might come about if the US government legislated oil companies differently: the jump of the case to Ecuador is an attempt to see if corporate laws might be stronger elsewhere, but the film is careful not to suggest that even in Correa's government property laws might be jettisoned. In *H₂Oil*, government hypocrisies are not linked to this or that party in power—such that an electoral shift would open up new possibilities—but to the operations of power around a commodity that will be excavated no matter what the health or environmental outcomes. *A Crude Awakening* is most directly about system failure: whether or not large social systems develop a greater awareness and more concerted direction about their energy futures, there is little sense that they can in fact meaningfully address the impacts of oil or manage to offset the looming civilizational crisis of oil ontologies. Existing systems have failed precisely by working all too well.[13]

On the evidence of this film, two axioms drive the social toward this "successful" system failure. The first is accumulation. Oil is far cheaper than drinking water or a Starbucks latte. As of August 14, 2018, West Texas Intermediate crude prices (US $66.76/barrel) put oil at $0.012 per ounce by comparison with $0.23 for a grande latte—a *nineteen*-fold difference (there are 5376 ounces in a barrel and 16 ounces in a latte).[14] It remains a primary commodity in global production and consumption systems that depend on an ever-increasing expansion of GDP as a measure of social wealth and of progress—the reason why economic growth trumps action on the climate in Alberta (and almost everywhere else). A second axiom operates at the level of the subject. One might ask: why do people work for Chevron? Or Petroecuador? Why do workers and technical experts flock to the spaces of oil production? It is unlikely that it is because there is strong support for the imperatives and initiatives of oil extraction and the economies it supports, but rather the need for work and fiscal security in an era in which the first axiom no longer encounters the impediments and strategies of a good (i.e., Keynesian) state. Lianne Lefsrud and Renate Meyer have studied the mechanism by which scientists involved in the Alberta oil fields explain to themselves their involvement in a process that they understand to have a climate impact: the availability of work enables a denial of scientific evidence even by scientists themselves.[15] There is a telling moment in H_2Oil in which spokespersons for an oil sands company are sent to address the concerns of the Dene First Nations. Their response to the criticisms by the Dene: they are only doing their jobs and not intending to hurt anyone. Their refusal to drink the local water suggests that they, too, suspect that the companies they represent are in fact causing damage to the environment and its human inhabitants. But they work for them anyway.

Antinomy?

The identification of a failure in the capacity of a broad range of social systems to address anything as serious as the crises generated by oil can lead only to one conclusion. If existing systems cannot address the problems these films bring up, everything has to change—new systems have to come into existence guided by new axioms. But how to move from here to there? There is an expected suggestion in each of these films that it is through education and the transmission of information to publics, which in turn will generate change through official and unofficial social and political networks, that politics "happens"—in other words, the gesture that politically committed documentaries tend to make toward the pedagogic effects of "seeing is believing." The opening segments of *A Crude Awakening* address bluntly the limits of knowledge about peak oil

(Congressman Bartlett: "Not one in fifty, not one in a hundred people in our country have an inkling of the potential problem we're facing"); part of the intent of the film is to transform this small minority into a majority. The addition of the didactic segments to the narrative of H_2Oil confirms director Shannon Walsh's hopes for the film to play a role in "educating a public who hadn't yet heard of the tar sands, and creating a context for further activism," and the ominous subtitle of *Crude* speaks to a similar desire to explain the "real price of oil."[16]

But even if the films never disavow this fundamental political aim, they recognize the complexity of the situations they encounter and represent, and they are cautious about the degree to which they are willing to figure their politics solely in relation to this pedagogic mode of knowledge transfer. These films frame two antinomies—first, that of the constitutive gap between knowledge and action, and second, between aesthetics and politics. While neither of these may be an antinomy in the strong sense of the term—they are not the same as the Kantian puzzle of the divide between natural causality and human free will, for instance—the suggestion of a blockage that seemingly no amount of conceptual thought or political activity looks likely to undo generates a genuine problem for knowledge and aesthetic practice. Antinomy here is meant to describe a stark social contradiction that emerges out of the messy activity of innumerable social systems. Generating an awareness of the structuring role of oil in civilizational processes, and so, too, of its obscene primacy over both human needs and ecological ones, produces on its own no resolution, even as it indicts the poverty of the present. As a genre, political documentaries like these three films might be seen as the invention of an imaginary solution to a social contradiction—the "imaginary" being the phantasmatic liberal public sphere it imagines into existence, that supposed space in which debate and discussion lead to a resolution that maximizes (say) individual freedoms within the demands of social necessity. These films gesture in this direction, but the substance they each address—oil—does not allow them to imagine that they do more than give evidence of the social contradiction produced by this sticky substance. The politico-aesthetic at work in these films gestures toward the possibility of audiences "doing something" because of the conditions in their world, while at the same time being unable to commit themselves fully to a belief that they can produce either an increase in knowledge or political action—less as the result of failure of political will, than due to a recognition of the constitutive nature of the world they produce and represent on film.

The productivity of antinomy is that it gestures to an overcoming that is present in the terms of the structuring division—one that requires only the right insight into the dynamic that produces the division to begin with. A crude, reductive (which is not to say unproductive) way in which to think about oil is to understand it as foundational to contemporary social form. The social contradiction is that the founding premise of society *as such* is draining away and cannot be replaced. Where exactly can one find synthesis in such a system, even if one were to undo the "enlightened false consciousness" that generates the gap between knowing and doing, evidence and action?[17] As a result of the demands of its subject matter, social contradiction in these films remains on their surface, whether they try to generate an imaginary resolution to it or not.

The failed sublime, or, scalar aesthetics

This final point emerges out of the previous one. A dominant aesthetic strategy in reference to oil is to emphasize scale. This is perhaps an obvious approach to a site like the Alberta oil sands, which are estimated to be the size of Florida and include numerous surface mining sites and vast tailings ponds that permit a direct visualization of environmental destruction.[18] But there are other ways to visualize and narrativize the scale of oil, too, including images of old drill sites on which derricks are clustered as tightly as bees in a hive, or the flow of traffic along freeways and through cities all over the planet. These images of cities and traffic are prominent in H_2Oil and *A Crude Awakening*, and identify the civilizational dependence on oil that will lead to crisis as its last dregs are used up; the former images of environmental impact, present in all three films, point to the astonishing degree to which human beings have remade a space as big as a planet, and continue to do so in ever more visible ways.

The use of scale in these documentaries is intended to add to knowledge and to generate an affective response. Is this not an appeal to the Kantian sublime in both of its aspects, the mathematical and the dynamical? Again, as with antinomy, the correlates are inexact: the palette of cities, however many different images of sped-up traffic we are shown, is not without limits, and the images of the oil sands are not of Nature but of its antithesis: "nature" after its encounter with humanity. Nevertheless, the gestures these films make toward representing oil through the visualization of scale do seem to have as their endpoint the same gesture as Kant's analysis of the sublime: to bring into cognition even that which seems to supersede and fall outside it. We are placed in awe of scale not so that we give up in the face of the vast existing infrastructure that depends on oil, or so that we concede to an ever-expanding tear in the face of the Earth (one now said to be visible from space), but so that it might provoke a closure of that gap between knowing and acting described above.

And yet this gap persists. Has the possibility of a politics through such scalar aesthetics collapsed? Throughout the history of film theory (starting with writers such as Jean Epstein and André Bazin), there is an insistence on the capacity of film to record what is otherwise inaccessible to vision, opening up reality to that quotidian experience which cannot help but miss reality's full ontological presence and depth. One should not disavow the capacity of documentary to bear witness to reality in just this way, at the level of both form and content; the sublime of oil culture that these films visualize does not readily appear to everyday experience, which is one of the reasons why the consequences of the end of oil are neither feared nor acted upon. One can see Kant's sublime as a domesticating process that renders what might well be alien to thought amenable to existing schema. The fact that the sublime fails is, then, not an issue, since its capacity to control and contain filmic images of traumatic scale in fact drains the latter of its effects, which is the exact opposite of what one might want. At the same time, however, abandoning oil to mathematical incomprehension or the terror of destroyed nature on a vast scale—the way in which a scalar aesthetics might be thought to do its work—seems to abandon thought to the inaction of what Slavoj Žižek has termed "cynical reason": awareness without action, even in the face of disaster, since we cannot possibly act on something that exceeds our comprehension.[19] In the end, what is incomprehensible is not the scale of our action on the world, but that our social world has as its foundation a substance demanded by our quotidian infrastructures, an input whose time has come, and soon will be gone. It is unclear what action one could take, even if one wanted to.

The Politics of Documentary in an Era of Scarcity

Natural resources can become depleted but human creativity is inextinguishable. I believe that once oil depletes, the genius of humankind will invent alternative sources of nutrition and fuel.

—Arman Medezuleyev, Baku oilfield operations manager,
A Crude Awakening

Michael Hardt's "Two Faces of Apocalypse: A Letter from Copenhagen" draws attention to the similarities in and differences of the politics of the anticapitalist and the environmental movements. Superficially, one might expect these two movements to be more similar than different, or even as occupying the exact same ground: a visual representation of this relationship would be less a Venn diagram in which there is a zone of overlapping concern (and so zones of exclusion, too), but one of two perfectly congruent sets that appear to be distinct only because each group spends more time in one part of the field than the other, thus misrecognizing the extent of their shared interests. Reflecting on his experience at COP 15 (the 2009 United Nations Climate Change Conference), however, Hardt recognizes that there are significant differences that would have to be addressed before each movement can operate fully in conjunction with the other.

Hardt identifies three antinomies that define and separate the anticapitalist and environmental movements (about the points of intersection—an opposition to property relations and their joint challenge to traditional measures of economic value—I will say no more). The first and defining one has to do with "a tendency . . . for discussions in the one domain [environmental movements] to be dominated by calls for preservation and limits, while the other is characterized by celebrations of limitless creative potential."[20] A second has to do with the question of knowledge. While "projects of autonomy and self-governance, as well as most struggles against social hierarchies, act on the assumption that everyone has access to the knowledge necessary for political action," Hardt writes,

> the basic facts of climate change—for example, the increasing proportion of CO_2 in the atmosphere and its effects—are highly scientific and abstract from our daily experiences. Projects of public pedagogy can help spread such scientific knowledge, but in contrast to the knowledge based in the experience of subordination, this is fundamentally an expert knowledge.[21]

The final antinomy grows out of a different relationship of each to time. For anticapitalist movements, radical change that would bring about the end of days is the opening to a new (and better) world. By contrast, for environmentalists, "the end of days is just the end," as the radical change that is likely on the horizon is one of "final catastrophe."[22]

The second antinomy is the one on which documentaries of the kind that I have been exploring here hope to do their work, either by translating expert knowledges into lay language or by producing accounts of damage to the environment that can be narrated and made visible, moving audiences from the specific (a film or a specific case) to the general (a confrontation with the issues facing the globe as a whole). When the subject matter is oil, it is impossible not to reflect on the terms of the first antinomy—that is, on limits, not only of Earth's environment but also of one specific element of it whose use has resulted in an assault on the environment even as it has contributed to or amplified the (apparent) limitlessness of human productive and imaginative capacity.

But it is the third antinomy that haunts documentaries on oil. The division Hardt points to in this third moment is, at least from one perspective, the least convincing. Does the end of days always already signal the effective destruction of the Earth's environment? Or can it not also speak to the possibility of a new world in which the antinomy between limitlessness and limit has been resolved (which is to say: what kind of revolution today could imagine that it has passed the end of days if it has not conceptualized what it means to live within limits?).

As long as it is figured in terms of climate change, the apocalyptical imagination of environmental movements continues to operate with an understanding of final catastrophe as temporally distant. When one thinks of catastrophe as the end of oil, however, the time horizon is pulled much closer, even as its politics are more difficult to cognitively map. For where should we place oil within this opposition of limit and limitlessness, the environment and the common? Oil is limited, and its use pulls closer that larger limit of the Earth's environment, of which it is simultaneously a part (limit) and an other (catastrophe) that the future would be better off without.

And what of the common and its limitlessness? A radical change to the present may well be precipitated by the evaporation of a commodity on which the common depends more than it might want to believe. "Fossil fuels helped create both the possibility of modern democracy and its limits";[23] given the problems of modern democracy, its evaporation alongside that of the energy inputs that helped fuel it might be welcome. But there is no guarantee that the new world on the other side of the end of oil will be one made in the image of revolutionary groups and their labors. The oil documentaries that I have explored here struggle with Hardt's antinomies and the political antinomies of crude aesthetics that I describe above, leaving open the question of how to resolve them (or even *if* they can be resolved), and refusing to offer solutions that would do little more than affirm that which they would seek to deny. Does this constitute a form of political success or failure? Or, perhaps their politics lie in the evidence they provide of the limit of what can be said about a socially ubiquitous substance that remains hidden from view—even today, and even in the process of bringing it to light.

NOTES

1. *Crude: The Real Price of Oil*, directed by Joe Berlinger (2009; New York: First Run Features, 2010), DVD; *A Crude Awakening: The Oil Crash*, directed by Basil Gelpke and Ray McCormack (2006; Zurich: Lava Productions AG, 2007), DVD; and H_2Oil, directed by Shannon Walsh (2009; Loaded Pictures, 2009), DVD.
2. See Gaines, "Political Mimesis."
3. Monani, "Energizing Environmental Activism?"
4. McLagan, "Introduction: Making Human Rights Claims Public," 192.
5. Wenzel, "Consumption for the Common Good?"
6. Jameson, *The Political Unconscious*, 79.

7.   Jameson, *The Political Unconscious*, 79.
8.   McLagan, "Introduction: Making Human Rights Claims Public," 191–95.
9.   Hardt, "Two Faces of Apocalypse."
10. See Burtynsky, *Oil*.
11. See chapter 4 of the present book.
12. See chapter 6 of the present book.
13. See Cazdyn and Szeman, *After Globalization*, 134–52.
14. Mark Shenk, "Oil Drops below $100, Gasoline Tumbles, on U. S. Supply Surge," *Bloomberg Businessweek*, May 11, 2011, https://www.bloomberg.com/news/articles/2011-05-11/crude-oil-falls-for-first-day-in-three-on-projected-gain-in-u-s-supplies (accessed August 17, 2018).
15. Lefsrud and Meyer, "Science or Science Fiction? Experts' Discursive Construction of Climate Change."
16. Ward, "The Future Is inside Your Sock: How People, through Documentaries, Can Make a Difference," NFB.ca Blog, May 10, 2011, http://blog.nfb.ca/blog/2011/05/10/the-future-is-inside-your-sock-how-people-through-documentaries-can-make-a-difference/ (accessed August 17, 2018).
17. See Sloterdijk, *Critique of Cynical Reason*.
18. One notable instance of the documentary use of scale is Peter Mettler's *Petropolis*, which consists entirely of aerial shots emphasizing the size and scope of Northern Alberta oil extraction. *Petropolis: Aerial Perspectives on the Alberta Tar Sands*, dir. Peter Mettler (2009; Greenpeace Canada), DVD.
19. See Žižek, *The Sublime Object of Ideology*.
20. Hardt, "Two Faces of Apocalypse," 271.
21. Hardt, "Two Faces of Apocalypse," 272–73.
22. Hardt, "Two Faces of Apocalypse," 273.
23. Mitchell, *Carbon Democracy*, 1.

CHAPTER 8

——

How to Know about Oil: Energy Epistemologies and Political Futures (2013)

How to know about oil: Is *how* the right question through which to frame an inquiry into the contemporary significance of oil? Is an epistemic question the right one? After all, do we not already know everything we need to about it—that this substance on which we depend for much of our energy generates geopolitical misadventures, environmental destruction, and (for some) massive profits?[1] Do we not already know that, because it is of necessity a limited resource, our dependence on it constitutes something like a civilizational category mistake—one that we are unlikely to rectify, not because we cannot identify the error, but because we are people who live in societies so saturated with the substance that we cannot imagine doing without it?[2] What could we possibly learn by thinking about how we know oil, as opposed to thinking about the ways in which we have lived with it and what we need to do to live without it?

There are two things implied in the "how" of the title of this essay. The first aims to draw attention to the multiple ways in which oil is framed as both problem and possibility, implying in turn multiple forms of being in relation to it. Oil is a physical substance—a thing identified by a concrete noun (like tree or chair) rather than an idea named by an abstract one (such as belief or identity). Moreover, condensed figures such as the smog-spewing freeway system to which oil's protean energetic utility has given birth stand as stark examples of its physical impact on the planet and our societies. Even so, oil only has such significance as it does for us as a result of the cultural narratives that shape our understanding of it. Despite being a concrete thing, oil animates and enables all manner of abstract categories, including freedom, mobility, growth, entrepreneurship, and the future in an essential way—an insight that recent cultural criticism is beginning to use to interrogate the energy-demanding structures and categories of modernity.

"How" is also meant to point to the fact that making oil part of our knowing—making it a key component of our investigations on whatever topic—changes how and what we know. Oil (and indeed, energy more generally) has almost always been seen as an external input into our socio-cultural systems and histories—that is, as a material resource squeezed into a social form that pre-exists it, rather than the other way around. We do not see it as giving shape to the social life that it fuels. It is thus that we imagine that life as we know it can continue along in its absence or disappearance, simply through the introduction of new, alternative sources of energy. With enough political will and technological innovation, we have a strong tendency to believe that wind, solar, geothermal, and nuclear energy could generate the kilojoules we have come to expect from fossil fuels, and do so in a way that would change our energy inputs while retaining the quality and form of life that many (though far from all) now enjoy.[3]

What if oil is *fundamental* to the societies we have now? What if it shapes them in every possible way and at every possible level, from the scale of our populations to the nature of our built infrastructure, from the objects we have ready to hand to our agricultural and food systems, and from the possibility of movement and travel to *expectations* of the capacity to move and interact, not to mention the plastics used to encase our smartphones and other high-tech devices? How we know about oil at the present moment tends to undervalue its impact and significance as a condition of possibility of modernity and of the full development of capitalism. If we insist on understanding modernity as an *oil* modernity and of capitalism as an *oil* capitalism, this cannot help but force us to reconsider how we understand both, as well as the ways in which left politics have been shaped over the past two centuries in response to the conditions produced by modern capitalism. In "The Fragment on Machines" in *The Grundrisse*, Karl Marx famously imagines a world in which technology has advanced to such a point that human labor is no longer required[4]—a fantasy of a laborless world that has animated some aspects of contemporary left thought as well. Even were we to achieve a world without work, the machines would still need oil to operate them, and the social systems and infrastructures (suburbs, highways, and the entertainment systems organized around them) inhabited by the laborers now free to do what they want would be ones that were brought into existence by oil and would need it to make it all operate.

As a contribution to the growing literature on oil and energy in the humanities—a flood of work that includes such recent texts as Matthew Huber's *Lifeblood: Oil, Freedom, and the Forces of Capital* (2013) and Stephanie LeMenager's *Living Oil: Petroleum and Culture in the American Century* (2013)—I want to explore what we might learn from three attempts to probe the consequences of how we know oil and how we might make oil a more conceptually powerful part of our knowing. The first of these attempts constitutes a reframing of the history of left politics in relation to changes in dominant forms of energy, specifically as this is explored in Timothy Mitchell's *Carbon Democracy: Political Power in the Age of Oil* (2011). The second—Edward Burtynsky's well-known and widely exhibited photo-series, *Oil* (2011)—narrates and names the social significance of oil through experiments in visual form. Finally, I will consider the struggle over the representation of the Alberta oil sands in public and political debate and discussion, looking here, too, at what happens when oil circulates as a contested cultural narrative as opposed to being merely a physical entity about which there is little dispute or debate.

I recognize that the three cases I have chosen to look at how oil is made part of our knowing are not of the same kind, nor do they operate at the same social level. An academic text, a photo-exhibit, and an ongoing political and media campaign about oil all work in different registers, involve distinct (if sometimes overlapping) communities, and have diverse ends in mind. Even so, I think that there is value to exploring these distinct interventions into our energy epistemologies. I want to do so to see what these attempts to know oil differently than we had known it—that is, as an essential component of social, cultural, and political form, and not just the caloric stuff that happened to propel modernity—might tell us about a politics appropriate to our petrocultures.

We stand at a moment when there is broad understanding and awareness of the need to make a transition from a global society based around non-renewable forms of energy to renewables. One of the reasons that there have been interventions into how we know oil and energy from multiple directions—history in the case of Mitchell, the visual in Burtynsky, and (broadly speaking) the political with respect to the oil sands—is that public knowledge about the environmental repercussions of oil usage, or of the consequences of its necessarily limited supply (however many years we may still be away from peak oil, a peak will come), seem to have generated limited political and social response, and nothing on the scale or with the speed required. Vaclav Smil has pointed out that "lessons of the past energy transitions may not be particularly useful for appraising and handicapping the coming energy transition because it will be exceedingly difficult to restructure the modern high-energy industrial and postindustrial civilization on the basis of non-fossil—that is, overwhelmingly renewable—fuels and flows."[5] I argue that these attempts to narrate new ways to know oil have lessons for a left politics committed to an energy transition that would both ameliorate environmental concerns and enable greater social justice.

The politics, presumptions, and implications of each of these ways of knowing oil—knowing it in order to understand just what it has meant for us moderns—varies, of course, and this is partly the point. Taken together, however, they point to important barriers to action and thresholds of possibility that we need to consider as we work against the overwhelming media and political promotion of oil as a benign force for good, to say nothing of the weight of quotidian comfort of our societies. These various *hows* draw attention to the compelling political openings that emerge once we accept and understand the ways that oil and energy animate our cultural narratives; they point, too, to the very real challenges and difficulties of trying to produce a different way of being in relation to a source of energy that has produced the societies we inhabit and has made us the subjects we are.

Alternative Histories

It is no exaggeration to suggest that the twentieth century would not have been the same without oil—a source of energy easy to store and transport, with a huge energy output per unit of fuel, and a source that forms the basis of all manner of other substances without which it is hard to imagine life on the planet today. Histories of the century that are alert to the significance of energy inevitably provide a vision of the recent past in which the presence of oil is among the central forces shaping human life—if not *the* single *ur*-force to which all other narratives can be connected. For example, J.R. McNeill's environmental history of the twentieth century, *Something New under the Sun* (2000), quickly identifies the capacities, technologies, and infrastructures enabled by fossil fuels to be the single most significant factor in the massive expansion of population over the century, which in turn generates the even larger increases in water consumption, carbon dioxide production, and more. The figures are staggering: a four-fold increase in world population, a 17-fold increase in carbon dioxide, and a 40-fold increase in industrial output—just to begin with.[6]

Daniel Yergin's Pulitzer Prize–winning commodity biography, *The Prize: The Epic Quest for Oil, Money, and Power* (2008), also places oil at the heart of human activity since its discovery for industrial uses in the late nineteenth century in Pennsylvania. One could pick almost any aspect of Yergin's book to make the case for the historical significance of oil, especially for the shape of economic and geopolitical history. In Yergin's account, for instance, much of what passes for military strategy in World War II can be reduced to the ceaseless appetite of mechanized armies for oil. Japan and Germany began from positions of energy weakness: no oil on native soil. As a result, the drive of the Germans to Russia and North Africa, and of Japan to Southeast Asia, was motivated by the need for energy to keep their militaries and economies on the move, as much as they were by popular-national narratives gone terribly awry. Of Pearl Harbor, Yergin writes that "the primary target of this huge campaign remained the oil fields of the East Indies";[7] the attack on the United States was carried out in order to protect the Japanese flank and to safeguard tanker routes to the home island from Borneo and Sumatra. At their worst, such oil histories can be reductive in a bad sense, seeing crude as always and everywhere the disease that generates the symptom called history, with its attendant traumas, dislocations, and crises. At the same time, an attention to the importance of oil contributes an essential and all-too-often missing element of our social and political narratives—the way in which energy, objects, and infrastructure exert demands on and shape human actions and decisions, giving form to the character and nature of political, social, and cultural systems.[8]

Timothy Mitchell's *Carbon Democracy* offers a powerful renarration of the politics of the petrocarbon era that is alert in just this way to the material significance of oil in shaping capacity and possibility, from the form taken by local struggles over oil to the shape of twentieth-century geopolitical conflict. Mitchell's book is not about democracy and oil, but about democracy *as* oil; Mitchell's aim is to show that "carbon-energy and modern democratic politics were tied intricately together."[9] The democracy to which he refers is "a mode of governing populations that employs popular consent as a means of limiting claims for greater equality and justice by dividing up the common world."[10] By reading modern democracy as oil—as made possible and enabled by oil in a fundamental, material way—Mitchell creates an alternative history that produces all manner of compelling conceptual openings for left and environmental politics.

There are two points from Mitchell's book to which I want to draw particular attention. The first is his remarkable account of the political effects of the emergence of the use of coal as a source of energy on a broad scale. One of the social transformations produced by coal was that for the first time the vast majority of people in industrialized countries became dependent on energy produced by others. The production of coal at specific sites across northern Europe that then had to be channeled to other sites along narrow railway corridors, with specialized groups of workers operating in large numbers at both ends, generated the material conditions for a form of political agency that could be asserted through the disruption of energy flow: "The rise of mass democracy is often attributed to the emergence of new forms of political consciousness. . . . What was missing was not consciousness, not a repertoire of demands, but an effective way of forcing the powerful to listen to those demands."[11] The ability of workers to disrupt energy flow effectively and immediately through mass strikes or sabotage gave their political demands special force, and led to major gains for workers between the 1880s and the interwar decades, while also supporting the development of workers' consciousness of their social circumstances. For Mitchell, the switch to oil from coal as the primary energy source for the Global North from the 1920s onward was a major factor in impeding the demands of labor and constituted the basis for a form of governmentality that managed the struggle for democracy. The production of oil requires fewer workers than coal in relation to the amount of energy produced; laborers remain above ground in the sight of managers; and from the 1920s "60 to 80 percent of world oil production was exported," which made it difficult to affect supply through strikes.[12] Mitchell is blunt in his claim: the mass politics that emerged alongside coal was defeated by the rise of fossil-fuel networks that made mass action more difficult, and changed the conditions within which class struggle took place.

The discourse of economics has played a key role in the system of democratic governmentality that Mitchell explores. Here, too, oil plays an essential, if hitherto unrecognized, role. Mitchell argues that the *economy* as an object did not exist in its current form prior to World War II. Before that time, the term *economy* referred to a *process* and not a *thing* whose management was to become the central task of governments and of a cohort of specialists who would produce knowledge about it. Nineteenth-century political economy concentrates on the "prudent management of resources applied especially to the resource that had made industrial civilization possible": that is, to coal.[13] This is an economy understood in terms of limits and scarcity. The shift from coal to oil enables a significant change in how the economy is conceptualized and governed. Nature is now removed from economic calculations. In place of natural resources and energy flows, economics becomes the measurement of money, and the economy transforms into a measurement of "the sum of all the moments at which money changed hands."[14] Mitchell argues that "the conceptualisation of the economy as a process of monetary circulation defined the main feature of the new object: it could expand without getting physically bigger."[15]

While it might not always have been the largest source of fuel (as measured by sheer volume used or energy produced), oil was nonetheless the twentieth century's dominant energy source. It remains the fuel source around which human life is configured: city spaces and infrastructures, global trade systems, forms of social experience (from mobility to individualism), and so on. And as the dominant form of energy, oil enables the idea of the economy as an object able to grow without limit in two ways.[16] First, because of its continuous decline in price (adjusted for inflation) over much of the century, the cost of energy was thought to have little bearing on economic activity or decision-making. Energy now appeared virtually free within overall calculations of the economy. Second, the apparent abundance of oil and the ability to move it wherever needed made it possible to treat it as inexhaustible. Mitchell concludes, "Democratic politics developed, thanks to oil, with a particular orientation towards the future: the future was a limitless horizon of growth. This horizon was not some natural reflection of a time of plenty; it was the result of a particular way of organising expert knowledge and its objects, in terms of a novel world called 'the economy.'"[17]

Left politics on the one hand, the economy on the other—the first impeded by the appearance of oil, the second fueled by it. Mitchell provides us with a shift in how we know oil that produces new possibilities for how we might act in relation to it. The connection between the apparent limitlessness of energy and our expectations of an economy that must, of necessity, continue to grow at all costs offers insight into the struggles faced by countries after the 2008 market crash. For economist Jeff Rubin, for instance, the difficulty of economic recovery can be put down to a single factor: the high price of oil, which at US $90-$100 per barrel by 2011 was close to 50-fold its average price during the period of capitalism's great expansionary phase from the 1920s to the 1970s, when energy appeared to be virtually free (as of August 2018, it remains expensive by historical standards, if far cheaper than its July 2017 peak of US $147.27).[18] How to imagine either stable liberal capitalist democracies or to consider alternative social formations requires us to consider carefully the role played by energy in shaping both their material realities and social imaginaries—the expectations for expansion and growth that oil has set up, which is proving challenging for the left to conceptualize otherwise in a concrete way in a world with ever-expanding populations and middle-class desires.

As for the politics of street actions and general strikes: while it would be reductive in the extreme to suggest that the switch from coal to oil as a dominant source of energy had long ago eliminated or reduced the efficacy of labor actions, it is worth taking seriously the connection between political action and energy. The difficulty of impeding the operations of capital today in anything like the way that may have been possible at mine sites is clear: major pipelines circumvent political hot spots and/or are buried beneath the ground;[19] and with the exception of environmental movements that draw attention to the causes and consequences of oil extraction and usage, almost nowhere is left political action organized directly in relation to energy in the way it once was.[20] While the decision by the global Occupy movement *not* to issue specific demands or to insist on any direct outcomes as a result of their street occupations has been viewed by some analysts as a rejection of the whole of late capitalist democracy and its presumptions, it can also be viewed as the last gasp of a form of protest politics that—understood from the vantage point of oil and energy—has, alas, long been outmoded.[21]

Mitchell's *Carbon Democracy* insists on the significance of oil to political form and possibility; in doing so, in naming just how deeply energy form shapes political form, he appears to rule out some of the forms of concrete action on which the left has traditionally relied to communicate to publics. At the same time, oil fuels the fantasies of limitless expansion in a way that has proved difficult to challenge or counteract, whether or not such growth is sustainable in the long run. How then might one explain or explore the political significance of oil to publics, especially given the need for large-scale, rapid energy transition? Might aesthetics succeed where street protests fail?

Aesthetics and Politics

Near the end of *The Long Emergency*, James Howard Kuntsler makes the claim that

The collective imagination of the public cannot process the notion
of a non-growth economy, even though the limits to growth are
visible all around us in everything from the paved-over suburban
landscapes, to the steeply rising gas prices, to played out aquifers, to
the death of the Atlantic cod fishery. We are not capable of conceiving another economic way. We are hostages to our own system.[22]

Such doubts about the capacity for the radical change necessary to
produce a new mode of social and economic organization are today all
too common. In his groundbreaking 1992 essay "Petrofiction," Amitav
Ghosh argues that one of the reasons for an absence of fictions about
oil and the social and political encounters this substance has produced
(Americans and Saudis, Canadians and First Nations, and so on) is
because it is "a Problem that can be written about only in the language
of Solutions"—a subject better suited to the binaries of public policy
or political science than to the language of aesthetics that deals with
contradiction and irresolution.[23]

Despite such worries and cautions, over the past decade there has
been a growing body of work in art and literature committed to naming
and explaining oil with the aim of producing changes in our social
imaginary.[24] One such project is Edward Burtynsky's 2009 exhibition
of photographs, *Oil*, which is made up of both new and old images
addressing the topic of oil from every possible angle. In Burtynsky's
characteristic style, which emphasizes scale, number, and hidden realities, these photos prompt shock and awe in the face of the visual
representation of the sheer size of the sites of oil extraction, the varied
infrastructures that enable it to course through the veins of global society, and the brute evidence of its physical, environmental, and social
consequences.

Burtynsky describes *Oil* as the outcome of an "oil epiphany" he had
in 1997: "It occurred to me that the vast, human-altered landscapes
that I pursued and photographed for over twenty years," he writes,
"were only made possible by the discovery of oil and the mechanical
advantage of the internal combustion engine. . . . These images can be
seen as notations by one artist contemplating the world as it is made
possible through this vital energy resource and the cumulative effects

of industrial evolution."[25] *Oil* is divided into three sections intended to document the life cycle of the substance, passing from "Extraction and Refinement" to "Transportation and Motor Culture" to "The End of Oil." The photos that make up each section are heterogeneous in theme and content, and photographed at numerous locations around the world. "Extraction and Refinement" includes images of older oil fields in the California desert, which are jam-packed with drill rigs and pumpjacks; the expansive oil sands extraction sites and tailings ponds surrounding Fort McMurray, Alberta; and the complex, visually dynamic twists and turns of refinery structures in Ontario, Newfoundland, and Texas. "Transportation and Motor Culture" begins with a series of Escher-like images of enormous highway interchanges, before transporting us to massive car import lots in the United States and China, and ending in photos of sites at which people accumulate around the fantasy of driving, as in the biker and trucker jamborees held in Sturgis, South Dakota, and Walcott, Iowa.

If the photos in the first two sections draw our attention to the apparatuses and infrastructures that produce and are produced by oil, from sites of extraction largely hidden from view to the quotidian landscape of highways and car lots, the final series, "The End of Oil," probes the consequences of oil society, especially through the detritus that it leaves behind. The multiple images of the ancient oilfields of Baku, as well as of gigantic graveyards of cars, helicopters, planes, jet engines, tires, and oil drums, are concluded with a sequence of photos on which Burtynsky first made his fame: the ship-breaking yards of Chittagong, Bangladesh, where nineteenth-century labor meets twentieth-century garbage through the mechanism of twenty-first-century off-shoring of multinational capitalism's expenses and responsibilities.

Oil is a photo-narrative—an attempt to tell a story through images. To visualize the world as it is due to oil, Burtynsky pieces this narrative together out of his existing large body of images in an effort to produce a tale that might generate in its viewers the same oil epiphany that prompted their production. One of the questions *Oil* raises is not just whether it succeeds in its political and pedagogic aim—too blunt a question to be posed to such a varied and vibrant set of images—but what we are to make of the visual mechanisms that Burtynsky employs in his photos and their capacity to name the central place of oil in our social imaginaries and ontologies.

The impulse of documentary photography with political aims is to engage in a form of didactic exposé—to introduce to vision otherwise hidden practices or spaces that we should know about, but do not, either because we do not want to or because we are not meant to. Burtynsky's images retain some of this impulse; but there is more going on. His attention to the spectacle of scale and the elevated vantage point from which his images are taken simultaneously exemplifies and critiques the enduring fantasy of Enlightenment knowledge. The god's eye perspective produces the enormity everywhere on display—a form of knowledge that makes it possible to see the outcome of petro-societies, but that is also able to create those economic and social systems that are able to leave signs of human activity on a planetary scale. Burtynsky's deserts are filled to the brim with cars and planes, and his images of garbage dumps—on a similarly otherworldly scale—track the detritus left behind when each of these is junked.

Epiphany means to understand the familiar "how" in some new way. In another register, it can mean that one finally comes to understand that one *does not* understand, or cannot possibly understand, what humanity hath wrought to the planet as a result of oil. The feeling one gets in moving through Burtynsky's photo-narrative of oil from birth to death is more the latter than the former—the dissipation of knowledge as opposed to its expansion. This is to his credit: the painful and beautiful images on display in *Oil* never stoop to render oil manageable, not even fully graspable, except as a dimension of contemporary social life from which we can no longer hide. Using Ghosh's terms, there is no Solution posed to the Problem on display, except perhaps to suggest that the accepted language of Solutions is always already part of the system that made the Problem identifiable as such. Mitchell points out that since there is no way to distinguish between beneficial and harmful growth, "the increased expenditure required to deal with the damage caused by fossil fuels appeared as an addition rather than an impediment of growth."[26] All of the images collected in *Oil* are images of growth, including the garbage graveyards and ship-breaking beaches.

Oil confirms Kunstler's worries, but it does so in a way that might yet generate the capacity for new social imaginaries.

One of the major difficulties faced by any aesthetic encounter with oil is the apparent capacity for the substance to absorb all critique, in much the same way that it absorbs light. In his critical reaction to Leo Lania's play *Konjunktur* (1928), which deals with the effects of an oil strike in Albania, Bertolt Brecht remarked that "petroleum resists the five-act form."[27] When novelists or visual artists decide to focus on oil—its environmental impact, the nature of the society that it fuels, the folly of depending on a finite energy source—it is because they wish to inform and to unsettle quotidian beliefs and behaviors, thereby activating a response in their readers and viewers with respect to oil. As Brecht intuited, however, the unique position of oil at the heart of contemporary society troubles the always-uneasy relationship between aesthetics and politics. The serious consequences of petro-societies demand a major—and speedy—collective response to a social problem almost without precedent. Whether this can be accomplished through aesthetic or cultural means, or whether the social significance of oil means that we occupy a novel situation for critical aesthetic practice with respect to society and the environment, remains an open question. What work such as Burtynsky's points to is how difficult it is for aesthetics to generate the kind and level of understanding required to produce desired social or political outcomes. Didactic exposé of the kind enacted by so many recent documentaries seems to do little but confirm the gap between knowledge and action that is a structural condition of modern life. If Brecht is right, and oil contains aesthetics, it may be that the only way to deal with the substance is by not taking it head on, but by trying instead to make more fully sensible the shape and form of the world to which oil gave birth.

Ethical Oil?

Making oil a central part of our histories introduces genuinely new insight into the shape those histories have taken and the politics that might be appropriate to the current moment; considering the ways in which oil is framed and named in aesthetics allows us to understand whether we can know it differently through art in a way that might allow us to reposition it in our daily practices and so create new social imaginaries. These two ways of knowing oil might be interesting and important, but also might seem to be less significant than the place in which a daily war is being waged about how we should—or can—think about oil: the media and the political sphere. This is perhaps especially true in Canada. There is a concerted representational struggle being carried out, on multiple fronts, nationally and internationally, over the status and significance of the Alberta oil sands, a place once of interest only to geologists but now part of everyday debate and discussion. With the reclassification in 2002 of the oil sands as part of the country's "proven reserves," Canada's oil stock jumped from 5 billion to 180 billion barrels, making it the second highest in the world after Saudi Arabia. As of 2018, oil and gas represented 40 percent of total Canadian exports—more than double what they were in 1995 (16.5 percent)[28]— and according to Natural Resources Canada, in 2010 the energy sector accounted for 6.8 percent of Canada's total gross domestic product.[29] Although Canadians are used to seeing themselves as "hewers of wood and drawers of water," this is no longer seen as curse but blessing.[30] As the relative importance of manufacturing declines, and as population shifts away from the East Coast and Central Canada, for both economic and demographic reasons political power and influence has also moved out West.

The development of the oil sands is opposed by environmental and First Nations groups, who have vocally and effectively drawn attention to the ecological trauma inflicted by the processes used in bitumen extraction. Opponents include not only those actors and environmentalists who have picketed the White House to try to influence the decision-making process with respect to approval of the Keystone XL pipeline, but also the opposition New Democratic Party (NDP). NDP leader Thomas Mulcair has come out as a forceful opponent of the oil sands (not so Justin Trudeau, the leader of the Liberal Party, who is a proponent of the energy industry and of the Keystone XL pipeline). In Mulcair's view, the oil sands have created a Canadian version of the "Dutch disease"—a decline in manufacturing linked to the artificial inflation of the value of the Canadian dollar. In British Columbia, former

NDP leader Adrian Dix put together a legal team, led by a prominent specialist in environmental, Aboriginal, and resource law, to help him devise a possible legal challenge to Enbridge's Northern Gateway pipeline. This pipeline would take bitumen from the oil sands to the West Coast for transport to China—what is seen by the federal government as an essential expansion of the market for Canadian oil, which is expected to grow from a current level of 1.2 mbl/day to 3.0 mbl/day by 2020. In both cases, NDP opposition is (or was) an explicitly political ploy: nationally, to gain support in the shaky economies of Quebec and Ontario; provincially, to win over an electorate that broadly supports the environment and is suspicious of the intentions of its resource-rich neighbors.

The federal government and industry groups have not stood idly by. In the 2012 budget, in addition to substantial cuts to a number of federal government ministries, the Conservatives announced a significant change to the environmental review process for new industrial projects. In addition to changes in process and policy, the government has ramped up its rhetoric in support of the oil sands on a number of fronts. Alarmingly, the federal government has begun to put pressure on environmental groups, suggesting that any who are involved in what might be deemed political advocacy could lose their status as charitable foundations. It has also suggested that the foreign contributions such organizations receive constitute the meddling of outsiders in internal affairs—a national-populist play somewhat uncommon in Canadian politics. This has already had an impact on some oil sands critics. Prominent scientist and environmentalist David Suzuki resigned from the board of the foundation that bears his name, while the charity ForestEthics has created a separate group—ForestEthics Advocacy—to make counterclaims to the government's representation of the oil sands and Canadian environmentalists.

For the claims, counterclaims, and rhetorical appeals of industry to function, they need to be seen as more than simply advocacy on the part of parties interested in profit at whatever cost to the planet. Industry groups have made a point of widely advertising their efforts to reclaim oil sands land and to act as responsible stewards of the environment; over the past several years, in addition to the other advertisements that cinema-going audiences have had to endure, they had to sit through advertisements by Cenovus (2011) that linked the company's new bitumen extraction technologies to the long history of Canadian innovation and its can-do attitude with respect to the scale of the country and the hostile cold of its winters (the punch line: "Canada: it's spelt [sic] with a 'can,' not a 'can't'"). Taking no chances, proponents of oil sands development have also flirted with a philosophical discourse about the oil sands—an ethics related to the form and shape of their development.

Ezra Levant's *Ethical Oil: The Case for Canada's Oil Sands* (2010) is not a genuine philosophical or theoretical text, nor is it an especially well-argued book, tending toward high-school debate-club style dismissals of opponents' positions through the deployment of what are at times relatively crude forms of rhetoric (a reliance on the identification of contradiction, expressions of startled surprise at the discovery of supposed hypocrisies, and so on). For all this, however, it articulates themes that lurk beneath many of the representational strategies of business and government concerning the oil sands: these make an appeal not only to science and economic necessity in the struggle for what Cymene Howe has termed "anthropocentric ecoauthority," but also to the good and the right course of action.[31] Whatever else ethics might be, they are intended to constitute the elaboration of the axioms and principles around which action and practice is shaped and governed, with respect to individuals, groups, or even non-human beings. Levant's book provides no such account; rather it engages in a broad attack on what he sees as the misguided ethics of those opposed to the development of the oil sands, alongside a concerted defense of the industry's practices. His arguments are anchored in a single core claim:

> Oil is an international commodity; if an oil-thirsty country such as China or the United States can't buy oil from one country, they'll buy it from another. So even if the oil sands were to completely shut down, the world wouldn't use one barrel less. It would just buy that oil from the oil sands' competitors: places like Saudi Arabia, Iran, Sudan, and Nigeria. . . . The question is not whether we should use oil sands oil instead of some perfect fantasy fuel that hasn't been invented yet. Until that miracle fuel is invented, the question is

whether we should use oil from the oil sands or oil from the other places in the world that pump it.[32]

Levant is unequivocal: by every possible ethical measure—whether human rights records, labor practices, or (most dubiously) environmental policies—Canada comes out ahead of its petro-competitors; he writes, for instance, "if Saudi Arabia didn't exist, it would take a science fiction writer in an apocalyptic mood to invent it."[33] The Panglossian verdict? Levant argues that "Canadian oil sands oil is the most ethical oil in the world, and the people who invest there, work there, and support the oil sands . . . are all, gradually, helping to make the world a more moral, humane, and better place."[34] This ethics of necessity, of the liberal "good guys" providing the energy to a world in desperate need of the stuff, is a way of knowing oil that absorbs any lingering anxieties and worries about the addition of Canada to a family of resource superpowers distinguished mainly by their dubious policies and politics.

It is clear what is at stake for government and industry in the multiple fronts on which they are waging the war of rhetoric over the growth and expansion of the Canadian oil industry. There is a great deal of money to be made in the oil sands. Alberta is crisscrossed with pipelines, and multiple pipelines already make their way across the border to the United States and over the Rocky Mountains to the Pacific Coast. The current struggle is to expand this infrastructure not only to accommodate new supply, but also to allow Western Canada Select Crude to trade at a price closer to world prices than its current discount rate.[35] Is this expansionary activity helping to make the world a better place? Such a conclusion can only be drawn from a too-easy acceptance of what already is: that we live in societies whose form and character depend entirely on oil. Of course, it is just this hypostatization of the political and social form and possibility that is challenged by those critics of the oil sands whom Levant and others on the political right critique.

That this sets the bar high for opponents to petro-societies does not invalidate their criticisms of the short- and long-term consequences of oil extraction and use. To Levant's logic of *either/or*—*either* the oil sands *or* something much worse, either (to put it plainly) democracy or fundamentalism—the proper response is to challenge the terms of the decision itself. In his response to the Bush administration's call for the US public to pick sides after 9/11, philosopher Slavoj Žižek famously argues that "What is problematic in the way that the ruling ideology imposes this choice on us is not 'fundamentalism' but, rather, *democracy itself*: as if the only alternative to 'fundamentalism' is the political system of liberal parliamentary democracy"[36]—or as if the only alternative to unethical oil is a supposedly more ethical one. One can reject both in favor of a possibility that exceeds and escapes the necessity of the given in favor of some third term yet to be named.

The parallel between politics and oil does not work, however; here we do reach an impasse in "how" we might know oil that is at once political and theoretical. What is unethical about the oil that Levant contrasts to the oil sands is in the main that it is nationalized and not owned by private companies. State-owned oil companies control 73 percent of the world's reserves; significantly, of the remainder, which is to say, privately owned oil, more than 50 percent is located in the Alberta oil sands.[37] For Levant, betraying the law of property always already means conceding defeat to the fundamentalisms of oil states such as Saudi Arabia, Iran, Sudan, and Venezuela. Whether oil is ethical or unethical, whether it might not be possible to have a national oil company that is not fundamentalist (Statoil, Norway's oil company, is mentioned only in passing by Levant), what persists amid all the rhetoric is the necessity of oil itself.

In a recent poll, two thirds of Canadians reported that they believed that the country could increase its oil and gas production without generating any further damage to the environment.[38] Wishful thinking in the extreme? Or the consequence of no better alternative to what is? The representational struggles that are taking place over the oil sands pit the language of marketing against the discourse of finitude. Government and industry want to make sure that the oil sands continue to be in demand in the way that one of Canada's other major exports— Blackberries—no longer are. To this, in the absence of a miracle fuel, the left can only offer dire warnings. James Hansen, head of NASA's Godard Institute for Space Studies and one of the people credited for first raising alarm about climate change, suggested in the *New York Times* that to exploit the oil sands fully would place civilization at risk.

The fantasy of greening something black has proven to be the more attractive vision not only because it holds out the possibility of amelioration as opposed to transformation, but because no real alternative has otherwise been proffered.[39] The only thing that the left can offer in its opposition to the oil sands, and indeed, to the use of oil more generally, is that we stop using it. Although it might be a challenge to do so, the better option would be to try to figure out how to generate ideas of how a world of seven billion people might use oil to different ends and for different purposes, and perhaps to a lesser extent than recent projections of the continued expansion of fossil-fuel use into mid-century.[40] Rejection of the resource constitutes no politics at all, hardly rising to the challenge of creating new ideas for how (and why) we might be able to live without oil.

Conclusion: How to Know about Oil

There are better and worse ways to know about oil. The large circuits of power and politics in which oil is intertwined were brought into focus during the 2003 Iraq War. During the global anti-war demonstrations that took place in February of that year, "No Blood for Oil!" signs identified petroleum as the true imperative of US military action in the country. In their analysis of the US invasions of Afghanistan and Iraq, the collective Retort argues that the connection between war and oil was in fact not borne out by facts on the ground. While the link established by publics between the exercise of military force and oil seems to suggest a form of political awakening and "*aspires* to be an economic explanation of history . . . it is really still locked inside a 'hero-and-villains' vision of social process. It revolves around the (malign) power of a single commodity, substituting the facticity of oil (and oil men) for the complex, partially *non*-factual imperatives of capital accumulation."[41] For Retort, oil must be seen as part of a larger capitalist system engaged in a form of ongoing primitive accumulation. However important a commodity it might be—oil is one of a handful of strategic commodities that remain "the motors of production, the ultimate hard currency of exchange"[42]—for Retort a narrow focus on oil identifies the wrong cause for the military forms of neoliberalism that have emerged in the first decade of the twenty-first century, which are driven by the necessities of accumulation and the problems of overproduction and under-consumption that have threatened capitalism since it entered the monopoly stage of production.

How else might we know oil besides this emphasis on system? The primary discourse in which oil is named and described is in relation to the environment. Here, too, oil has been treated as one aspect of a larger problem—whether conceptually, within philosophies that identify alienation from nature as one dimension of social alienation,[43] or in relation to those material and economic processes that generate the human impact on the planet. The incorporation of ecology into left political theory has to date been relatively (one might also say surprisingly) easy. At its core, capitalism is a system that generates private riches by using up public wealth—that is, natural elements such as water, air, soil, and wood—and by removing workers from the Earth as the site of the production of their own lives. According to John Bellamy Foster, who is probably the most well-known theorist linking Marxism and ecology, the real moment of ecological crisis arises with the monopoly stage of capitalism. The problem of having to absorb surplus production was "resolved" through the creation of the consumer society we continue to enjoy and endure.[44] This society is one characterized by enormous waste and inefficiencies in the generation of the necessary profit from production. For Bellamy Foster and other Marxists, it is the ever-widening gap between socially necessary production and the massive superfluous production characteristic of contemporary capitalist society that transforms capitalism's fundamental imperative of shifting public into private wealth within a system that, due to its intensity and scale, threatens life on the planet as such.

Monopoly capital is enabled by oil in a more direct way than Retort might believe—which is not to say that oil was the direct cause of the war with Iraq. The expansion of capital could not have occurred in the absence of oil as a hegemonic form of energy. It is not only economists who have a blind spot with respect to the function of oil in contemporary economic and social systems and the objects and infrastructures they have produced over a century and a half of perpetual expansion. The elimination or reduction of wasteful, excessive production and consumption through a new social and political system no longer premised on consumer society *would* produce a socialism that had a much smaller ecological footprint. That socialism, however, would still require a significant amount of energy to fuel it, especially if it had as one of its goals the reduction or elimination of necessity in order to engender greater and greater social freedom. Labor is one of the key elements of this necessity; the possibility of freedom from labor through the increasing productive capacities of industrial systems has long fueled left imaginings of future social possibilities, from Marx through to mid-century imaginings of a gradually reduced work week to Italian *operaismo* and its theorization of the shift from the factory to the social factory. The political possibilities of the limitless capacities of creativity that accompany the hegemony of affective or creative labor, as theorized in different ways by Michael Hardt and Antonio Negri and Paolo Virno, are nonetheless still weighed down by the very real, material limits of the energy required to fuel the systems in which such creative, affective work is carried out.[45] It may be reductive to position oil as the *ur*-commodity that fuels two imaginaries—that of capitalism *and* socialism. It *would* be a mistake, however, to treat it as less significant than it in fact is: a substance fully deserving a prominent role in those alternative histories, aesthetic speculations on the future, and political struggles in the present that I have investigated here. The energy that makes modernity possible has, until recently, never been named, and its conceptual, as well as social and historical, significance never explained. As I hope I have shown in the various ways that we know oil—or need to know it, need to understand that our knowledge of it takes different forms with different repercussions—this is something that is essential to add to our calculations of future political possibilities, as well as to our understanding of the past within which our ideas of labor and capital were born and shaped into forms on which we still rely.

What if we were to renarrate left politics with the energy and capacities introduced by oil as a key element of the story? How might this force us to reimagine both anticapitalist and environmental politics? The insights offered by Mitchell about the significance of oil in shaping and enabling contemporary democratic governmentality tells us more about the directions taken by capitalism than the counterpoints that have hitherto been offered by the left. The civilizational possibilities introduced by oil are seductive and far easier to defend with representational fictions of petro-plentitude—which accord both with the specialized narratives of economics and with quotidian common sense—than with (still abstract) ideas and ideals of environmental devastation on the horizon, however close that horizon might be. Some representational openings might be generated by aesthetic interventions into oil imaginaries of the kind offered by Burtynsky's *Oil*, though most examples of oil art are more determinate and didactic—which is to say unrealistic—in their renderings of what petro-societies might be in the absence of petroleum. Even work as careful as Burtynsky's has to struggle with its capacity to intervene meaningfully at the level necessary to generate social and political change, whether due to the representational struggle over oil taking place in the media or the theoretical challenges that Jacques Rancière and others have made recently to assumptions guiding the concatenation of aesthetics and politics that have long fueled the energies of cultural critics.[46] The introduction of oil and energy would not invalidate left thinking, but make it more alert to the necessity of mass energy for the enormous social and infrastructural systems we inhabit and those we prophesize. It would also alert us to the dead end of any environmental discourse that continues to ally itself with economics (as in some variants of theories of sustainability)—a discourse that depends on oil being virtually "free"—and the need to create aesthetic *and* political interventions that oppose the narrative of endless growth with something more direct and more powerful than the ecological ethics on which we continue to depend.

Understanding how we know oil, and how we might or should know it, should make us alert to the very real challenges of naming, thinking, and changing the global society and social imaginaries that we have constituted around black gold, a substance that has given us the force to shape ourselves into what we are; we depend on oil in daily life, and even as we endure the consequences of having used so much of it, we one day will have to do without it. The insights into how we know oil that emerge out of work such as Mitchell's and Burtynsky's, and from the public struggles over the political significance of sites of oil extraction and use, as in the case of the Alberta oil sands, do not come close to addressing the challenge that Smil poses in his framing of the difficulties of transition from fossil-fuel to non-fossil-fuel societies. Thinking about the diverse and distinct ways in which we know oil and how we might come to know it differently highlights the necessity—and very real difficulty—of naming the problem and narrating the changes needed in a way that does not simply reinforce the inevitability of oil and the impossibility of the transitions we so desperately need. We are only at the beginning of the critical process of *really* knowing oil, of knowing it as fundamental to the determinations of our subjectivities and the shape of our social lives. Only by knowing oil can we start to understand fully what and who we might become without it—a task that needs to be at the heart of our political thinking today.

NOTES

1. There has been an explosion of books over the past decade that examine the implications of oil for geopolitics, the environment, and economics. Among the most prominent and influential of these, including research with specific reference to Canada, are Coll, *Private Empire*; Heinberg, *The Party's Over*; Heintzman and Solomon, *Fueling the Future*; Homer-Dixon, *The Carbon Shift*; Humphreys, Sachs, and Stiglitz, *Escaping the Resource Curse*; Klare, *Blood and Oil* and *Rising Powers, Shrinking Planet*; Marriott and Minio-Paluello, *The Oil Road*; Montgomery, *The Powers That Be*; Rubin, *The End of Growth*; Simmons, *Twilight in the Desert*; and Yergin, *The Quest*.

2. An increasing number of articles and books have endeavored to draw attention to the broad cultural and political implications of our societal dependence on oil. Such works offer re-narrations of history and speculative future projections of social changes and developments as a result of changes in energy use, and also explore (in a way similar to the project of this essay) the theoretical implications of attending to oil and energy. See Boyer, "Energopolitics and the Anthropology of Energy"; Burkett and Bellamy Foster, "Metabolism, Energy, and Entropy in Marx's Critique of Political Economy"; Knechtel, *Fuel*; Negarestani, *Cyclonopedia*; Nikiforuk, *The Energy of Slaves*; Stoekl, *Bataille's Peak* and "Unconventional Oil and the Gift of the Undulating Peak." There have also been intriguing efforts to understand how oil and energy shape feeling and belief. See Hitchcock, "Oil in an American Imaginary" and LeMenager, "Petro-Melancholia" and *Living Oil*.

3. For a discussion of the dominant public narratives of the future of fossil fuels, see chapter 4 of the present book, especially the section on "techno-utopianism."

4. Marx, *Grundrisse*, 704–6.

5. Smil, *Energy Transitions*, 105.

6. McNeill, *Something New Under the Sun*.

7. Yergin, *The Prize*, 326.

8. The study of the agency of objects is the focus of new materialist criticism, which attends to "efficacy of objects in excess of the human meanings, designs, or purposes

they express or serve." Bennett, *Vibrant Matter*, 20. A model of this kind of research in relation to oil can be found in Matthew Huber's *Lifeblood*, which addresses from multiple perspectives "the materiality of oil and how it shapes its 'system of provision'"—oil as a "specifically material aspect of the alienated—seemingly autonomous—power of capital over living labor" and as "a central energy resource shaping the forces of social reproduction, or what I call the real subsumption of life under capital." Huber, *Lifeblood*, xix.

9. Yergin, *The Quest*, 5.
10. Mitchell, *Carbon Democracy*, 9.
11. Mitchell, *Carbon Democracy*, 21.
12. Mitchell, *Carbon Democracy*, 37.
13. Mitchell, *Carbon Democracy*, 127.
14. Mitchell, *Carbon Democracy*, 136.
15. Mitchell, *Carbon Democracy*, 139.
16. To get some sense of the importance of fossil fuels to the shape and character of the twentieth century, consider Edward Renshaw's now half-century old study of changing energy inputs into the economy. Renshaw points out that "animals contributed 52.4 per cent of total work output in the United States in 1850; human workers, 12.6 per cent; wind, water, and fuel wood, 27.8 per cent; and fossil fuels, 6.8 per cent. In 1950, work animals are estimated to have contributed only 0.7 per cent of total work output; human workers, 0.9 per cent; wind, water, and fuel wood, 7.8 per cent; and fossil fuels, 90.8 per cent." Renshaw, "The Substitution of Inanimate Energy for Animal Power," 284. Recent studies have indicated that by 2020, coal will once again be the dominant source of energy in the world. Florence Tan, "Coal to Surpass Oil as Top Global Fuel by 2020: Report," *Financial Times*, Oct. 15, 2013.
17. Mitchell, *Carbon Democracy*, 143.
18. Rubin, *The End of Growth*.
19. For the difficulties of activating a politics around buried pipelines, see James Marriott and Mika Minio-Paluello's account of retracing the path of the BTC pipeline from source (in Azerbaijan) to mouth (in the Mediterranean Sea). Marriott and Minio-Paluello, *The Oil Road*.
20. Energy sabotage today invites a ferocious response from government security mechanisms. The Northern Alberta farmer, Wiebo Ludwig, repeatedly raised alarms to all levels of government about the health impacts of toxic sour gas from nearby wells on members of his family. When his entreaties were ignored and dismissed, he took matters into his own hands, destroying wells and pipelines in the area near his farm. He and his family are suspected of sabotaging wellheads and pipelines near their property more than 100 times. In 2000, Ludwig was sentenced to 28 months in jail. See Nikiforuk, *Saboteurs*.
21. Bernard Harcourt argues "Occupy Wall Street immediately fashioned a new form of political engagement, a new kind of politics. It is a form of political engagement that challenges our traditional political vocabulary, that ambiguates the grammar we use, that playfully distorts our very language of politics." Harcourt, "Political Disobedience," 46. Equally, however, one could view Occupy not as a form of radical refusal of an intervention into the existing terrain of rights and freedoms, but as

an example par excellence of the exhaustion of an existing form of political struggle—the street action; Occupy was likely both of these at the same time.

22. Kunstler, *The Long Emergency*, 193.

23. Ghosh, "Petrofiction," 139.

24. In the art world, oil and gas have become subjects that artists have begun to approach head-on through a variety of mediums and a range of perspectives. As two examples of a growing body of work, see Arns, ed., *The Oil Show*, an exhibition held at the Hartware MedienKunstVerein in Dortmund, Germany, and Ernst Logar's *Invisible Oil*, a catalog of a solo show staged at Peacock Visual Arts in Aberdeen, Scotland. A large number of documentaries about the politics of oil have also been produced since 2000. For an analysis of, respectively, art film and didactic documentaries on oil, see chapters 6 and 7 of the present book. I would like to thank Maria Whiteman for her help with this section.

25. Burtynsky, *Oil*, 2.

26. Mitchell, *Carbon Democracy*, 140.

27. Brecht, *Brecht on Theatre*, 29.

28. Dave Cooper, "West Coast Pipeline Key to Canada's Interests, Economist Says," *Edmonton Journal*, May 10, 2012, C1.

29. Natural Resources Canada, *Important Facts on Canada's Natural Resources*, 14.

30. The phrase comes from the Bible: "You are now under a curse: You will never cease to serve as woodcutters and water carriers for the house of my God" (Josh. 9:23; English Standard Version). The connection to Canada was famously made by Harold Innis in 1930.

31. Howe, "Anthropocenic Ecoauthority."

32. Levant, *Ethical Oil*, 6–7.

33. Levant, *Ethical Oil*, 15.

34. Levant, *Ethical Oil*, 234.

35. Reuters, "Canadian Crude Discount Expected to Stay Steep," *Financial Post*, Oct. 17, 2013.

36. Žižek, *Welcome to the Desert of the Real*, 3.

37. Mark Golden, "State-owned Oil Companies Increase Price Volatility and Pollution, but Rarely Get Used as Geopolitical Weapons, Says Stanford Researcher"; Canadian Association of Petroleum Producers, *Upstream Dialogue*.

38. Postmedia News, "Canadians Believe 'Green' Oil Development Possible: Poll," *Financial Post*, May 3, 2012, http://business.financialpost.com/2012/05/03/canadians -believe-green-oil-development-possible-poll/?lsa=d9667c02 (accessed August 17, 2018).

39. James Hansen, "Game Over for the Climate," *New York Times*, May 10, 2012, https://www.nytimes.com/2012/05/10/opinion/game-over-for-the-climate.html (accessed August 17, 2018).

40. Yadullah Hussain, "Fossil Fuels Reign, Even in 2050: Study," *National Post*, Oct. 18, 2013, FP7.

41. Retort, *Afflicted Powers*, 42.

42. Retort, *Afflicted Powers*, 39.

43. For an overview, see Biro, *Denaturalizing Ecological Politics*.

44. Bellamy Foster, "The Ecology of Marxian Political Economy."

45. See Hardt and Negri, *Empire* and Virno, *A Grammar of the Multitude*. For a

discussion of the antinomies that exist between the contemporary anticapitalist and environmental movements, see Hardt, "Two Faces of Apocalypse."

46. See Rancière, *The Emancipated Spectator*, and Raunig, *Art and Revolution*.

Chapter 9

Entrepreneurship as the New Common Sense (2015)

We have not emerged from the "iron cage" of the capitalist economy to which Weber referred. Rather, in some respects, it would have to be said that everyone is enjoined to construct their own.

—Pierre Dardot and Christian Laval, "The New Way of the World, Part I: Manufacturing the Neoliberal Subject"

In its May 2014 issue, *Wired* magazine featured an account of the trials and tribulations of Boomtrain, a Silicon Valley start-up that was struggling to find the funds to get started up. The Netflix style recommendation engine that Boomtrain was creating was of some interest to investors and bigger tech companies. However, these investors wanted to see the fruits of the firm's high-end labor before passing along serious money. In the increasingly cart-before-horse world of investment funding, Boomtrain was thus finding it difficult to keep afloat as it tried to cash in on the Internet boom. The plucky duo at the center of Boomtrain vowed to keep at the project nonetheless. After all, hard work, risk-taking, and the ferocious challenge of a crowded, competitive marketplace were a necessary part of the high-tech entrepreneurial adventure—a series of trials that, rather than constituting an impediment to their efforts, all but guaranteed success in the end.

Articles about entrepreneurial efforts in the tech world are the bread and butter of *Wired*, which gravitates to reports about the smarts of computer nerds and the savvy of money men, told through narratives that make it seem as if financial and social success is, in the main, inevitable in the new world of the devices and gadgets that increasingly mediate our lives. What made this article about Boomtrain distinctive is that it challenged the "fantasy that entrepreneurship—and, more broadly, creativity—can be systematized . . . that success in the startup game can be not only taught but rationalized, made predictable."[1] "The promise of professionalized entrepreneurship has had a particular allure in recent years," writes the article's author, Gideon Lewis-Kraus. "Starting a company has become the way for ambitious young people to do something that seems simultaneously careerist and heroic."[2] Yet, despite its warnings about the impossible conditions faced by tech innovators, Lewis-Kraus's cautionary tale does little to deflate the dream of entrepreneurial success currently circulating in the world. In the context of a magazine devoted as much to the cause of the entrepreneur as to reporting on the cool new gadgets and apps, the story of the struggles faced by Boomtrain comes across as an exception to a now general and increasingly widely accepted rule: the entrepreneur has become a model of how to be and behave, and not only in the world of business. Entrepreneurship has come to permeate our social imaginaries in a way that has quickly transformed its claims and demands on us from fantasy into reality. We are all entrepreneurs now, or, at a minimum, we all live in a world in which the unquestioned social value and legitimacy of entrepreneurship shapes public policy, social development, economic futures, and cultural beliefs and expectations.

The ideas and ideals of entrepreneurship constitute a new way of being and behaving in contemporary social life, the full force and social import of which are only now beginning to be felt. Its signs are all around us: from the millions of dollars expended by governments on programs and policies to support entrepreneurship to the flurry of new business books devoted to the subject;[3] from the global success of the reality TV show *Dragons' Den*[4] to Mattel's recent announcement of a line of Entrepreneur Barbie dolls;[5] and from the explosion of start-up incubators and accelerators[6] to the rise of programs and institutes devoted to entrepreneurship at universities around the world.[7] Even fields commonly thought to exist outside of the sphere of business and labor, such as artistic and cultural production, have been colonized by discourses of entrepreneurship.[8] Entrepreneurship exists in the twenty-first century as a commonsense way of navigating the inevitable, irreproachable, and apparently unchangeable reality of global capitalism.

The concept of entrepreneurship extends back to the eighteenth century, when economist Richard Cantillon famously described the term "entrepreneur" as a "bearer of risk."[9] This idea of the function of the entrepreneur, reinforced in the early twentieth century by economist Frank Knight and others, has remained the same for much of capitalist modernity. Entrepreneurship was understood as an important if minor element of capitalism—the site at which individuals and small groups would take chances on hitting the economic jackpot through the invention of new products, services, and means of distribution. These risky interventions would take place within an extant social landscape in which power was mobilized and organized via states and corporations in largely predictable ways. The exciting dice toss of entrepreneurship worked only to the extent that the general operations of capital were framed by the boring logics of standard economics. To some extent, it may seem as if entrepreneurship has remained a minor practice, especially given the continued presence of large corporations offering "fixed wages" (Cantillon's rubric for non-entrepreneurs) whose reach now extends across the globe. William Whyte's famous investigation of the postwar corporation, *The Organization Man*,[10] which explored the collectivist ethic that shaped US businesses such as Ford and General Electric, finds its equivalent today in texts detailing the operations of giant high-tech companies, including Facebook executive Sheryl Sandberg's bestselling corporate autobiography *Lean In: Women, Work, and the Will to Lead* and Dave Eggers's *The Circle*, a novel about the dangers of large e-corporations and the impact of their practices on democratic life and individual privacy. The content and ethos of corporate power might have changed substantially over the past half century, but its overall form and social function seem, at least from one perspective, to have changed less than we might think or hope.

And yet, even if one cannot gainsay the continued presence and power of corporations, what has changed is the status of the entrepreneur, who has been thrust to the center of the economic imaginary with breathless speed and insistence. The entrepreneur stands as an exemplary figure because, as Stuart Hall and Alan O'Shea characterized the commonsense importance of another contemporary structure (neoliberalism), it provides a basis for "frameworks of meaning with which to make sense of the world."[11] Entrepreneurship is a sticky idea around which contradictory and multiple constellations of other ideas coalesce; like many instances of common sense, this one sutures together certain (irresolvable) contradictions and challenges, making the existing situation seem natural, to-be-expected, and thus not only bearable but (in this case) anticipated and exciting. Rather than appear as a folkloric, inconsistent, and internally riven structure of everyday logic, common sense presents itself as the spontaneous realization of rational individuals, so that instead of being seen as a singular fluke, the entrepreneur is abstracted and universalized into a model for all citizens (indeed, a model that may have the potential to replace the citizen as such). As an aspect—perhaps the central aspect—of how we are coming to imagine the correctly functioning society, the figure of the entrepreneur is one around which political, economic, aesthetic, and educational structures have been and are still being reshaped: this more than anything else justifies its exemplary position at the outset of the twenty-first century—a position that is quickly coming to define the normal operations of markets and societies around the world.

No longer a minor figure at the margins of capital, slowly propelling it in new directions, nor just a harmless buzzword of the age, soon to be replaced by another when its glow starts to fade, the entrepreneur is the neoliberal subject par excellence—the perfect figure for a world in which the market has replaced society, and one whose idealization and legitimation in turn affirms the necessity and veracity of this epochal transition. The figure of the entrepreneur embodies the values and attributes that are celebrated as essential for the economy to operate smoothly and for the contemporary human being to flourish. When self-proclaimed "startup evangelist" Anna Vital claims on her blog that soon "everyone will have to become an entrepreneur," it is only in part because of the miraculous process through which they "create value out of nothing."[12] In a period when many workers in the Global North toil away at what David Graeber has termed "bullshit jobs"[13]—meaningless positions in administration and service that seem to exist "just for the sake of keeping us all working" rather than actually producing anything[14]—and older forms of work in productive industries have been automated into nonexistence, entrepreneurial engagement is held up as a way to both gain an income and give one's life meaning: simultaneously careerist and heroic, to return to Lewis-Kraus's Boomtrain article.

We are all entrepreneurs now; everyone will have to become an entrepreneur. If entrepreneurship has become common sense, it is in part because of the degree to which governments have become involved in the process of creating entrepreneurial subjects. In Canada, for instance, otherwise cash-strapped governments have freely committed millions of dollars to programs to support entrepreneurial endeavors, with the majority of programs targeted at youth.[15] The college and university system stands as one of the primary sites to which these funds have been directed. In Canada, an October 2013 report by the Council of Ontario Universities put it bluntly: "Entrepreneurship, upon which economists say economic growth depends, has moved from the margins to the mainstream of university education."[16] In the United States, courses of study in entrepreneurship are among the fastest growing programs at both undergraduate and graduate levels,[17] and universities have already come to be ranked according to the number of entrepreneurs and companies they produce as well as the level of capital raised by their students.[18] In entrepreneurial studies programs one does not dryly study the phenomenon of entrepreneurship—track its history, understand its function and role in contemporary capitalism, and figure out what entrepreneurs do. Rather, as the description of Northern Michigan University's (NMU) "Entrepreneurship Major" makes clear, these programs are explicitly designed to create new forms and modes of subjectivity:

> At NMU, we believe entrepreneurship is a mindset . . . a way of thinking, of acting, of engaging the world in a pursuit of new opportunities in the face of risk and uncertainty. Entrepreneurs are both dreamers and doers, market leaders, as well as market finders. For us, entrepreneurship is a full contact extreme sport. Our program is geared for you to learn a skill set that will increase your likelihood of success in an entrepreneurial setting or even a corporate setting in need of entrepreneurial thinking. In our program, you will learn by doing not just by reading from a book.[19]

Why are entrepreneurial subjects needed today? The language of risk and uncertainty that has always accompanied entrepreneurial activity has today become generalized. Everyone has to be an entrepreneur because in the absence of society—of the guarantees of formal and informal security and welfare once provided by community and state policies and programs—risk is a universal condition of existence.

There are two principal dimensions of today's universal risk and uncertainty. First, the disappearance of sites and spaces available for accumulation engenders a desperate need for state and capital to innovate their operations.[20] The risk for contemporary capital today is an outcome of *both* the need for change in a period of ever more intensive and extensive processes of accumulation *and* the potential failure to make the changes necessary to be able to take advantage of new sites and spaces of accumulation. Either changing or standing still may result in failure; success, on the other hand, has become far more difficult in a period marked by increasingly limited possibilities of growth. For late capitalism, entrepreneurial subjects are ideal ones, as they make minimal demands on the state while also working tirelessly to ferret out new possibilities for profit within a system whose logics have brought it treacherously close to collapse.[21] In this context, investments in the production of entrepreneurial subjects via the mechanisms of (for instance) post-secondary education are a small price to pay to address threats to the system from both stasis and change.

The second dimension of risk today exists for those subjects who inhabit contemporary capitalism. As a great many commentators have noted, contemporary life is lived in increasingly difficult and precarious circumstances.[22] Indeed, in *Precarious Life*, Judith Butler names our corporeal existence as *fundamentally* precarious—a state of vulnerability shared universally, if distributed unequally across the globe, and offset only by the practices of community and society that might help make life less nakedly violent and treacherous. Such ameliorative practices—social assurances and formal insurances, for example, which developed in many countries following World War II—have, at least since the 1980s, had their efficacy and efficiency impeded when they have not been dismantled altogether. Though some thinkers—most notably Giorgio Agamben—treat precarity as a mode through which power is mobilized and sovereignty consolidated via directed exclusions of and within populations, precarity has in fact become a universalized condition of contemporary experience due to the practices of the neoliberal state and global finance. Entrepreneurial subjects arise in response to this universal precarity: they are actors needed by states and capital alike to invent new forms and spaces of accumulation, but they also constitute a model of subjectivity appropriate to the uncertainties that attend contemporary capitalism. Instead of chaffing and complaining about the retreat of the state and the disappearance of society, or about their abandonment to the hostile environs of the contemporary labor market, entrepreneurs embrace the openings left behind by the retreat of the state as spaces wherein they can shape their own subjectivity with the greatest freedom imaginable. In this light, the disappearance of the state is not viewed as a consequence of a series of political decisions that favor the operations of capitalism and further the ends of an economic elite, but rather as a clearing of the way for a new condition of contemporary life linked to the historical victory of liberalism and the ubiquity of new communication technologies, and the fantasies of the unfettered individual freedoms that attend both. In a perverse way, the new programs of entrepreneurship appear to meet a demand that pre-existed them, and not vice versa; states appear to be doing only what their precarious subjects want them to in order to help them actualize their economic possibilities and social freedoms.

For the entrepreneurial subject, the suspension of society and its collectivist impulses has produced opportunities hitherto unavailable. For such subjects, the world is not replete with divisions of power and privilege that skew one's opportunities within it, predetermining possibilities through a game of social and economic fate. On the contrary, a society of entrepreneurial subjects confronts the world in a variant of the Rawlsian "original position," experienced now not as a thought experiment, but as reality.[23] As Dardot and Laval note, for the entrepreneurial subject, "the distribution of economic resources and social positions is exclusively regarded as the consequence of trajectories, successful or otherwise, of personal realization."[24] Entrepreneurship aims to level the playing field within societies, making success or failure no longer a condition of social differences or state power, but a matter of individual ability and desire. In the entrepreneurial imaginary, we all start on equal footing. The availability of new technologies and the accessibility of information about the general operations of the economy and society means that individuals can shape themselves through their own efforts into whatever they might want to become. The discourse of the entrepreneurial subject thus confuses formal and actual freedom; it consists of a strong belief that the formal freedoms we experience within contemporary capitalism in fact constitute an actual freedom that might challenge the limits of capitalism itself.[25]

We can see this misrecognition of the lived realities of contemporary society at work most obviously in the dreams championed by Silicon Valley, as in Brian Chesky's call for a "Shared City" or Peter Thiel's Seasteading Institute and Thiel Fellowships,[26] and in all kinds of so-called social entrepreneurship as well. In his "Shared City" manifesto, Chesky (founder of the hospitality start-up AirBnB and proponent of the sharing economy and collaborative consumption)[27] laments the loss of community in modern, urban societies, and calls for new forms of community fostered through the sharing economy via companies like AirBnB, Uber, Lyft, and others. Chesky hopes to "foster and strengthen community [and] bring back the idea of cities as villages," in which "people become micro-entrepreneurs, and local mom and pops flourish once again."[28] Thiel (cofounder of PayPal and a serial entrepreneur) cofounded the libertarian Seasteading Institute in 2008; the organization has as its goal "the creation of ocean city-states in order to advance humanity through innovative startup governments."[29] The Thiel Fellowship program was founded in 2011 to "rethink what it takes to succeed and improve the world" by providing up to twenty young people under twenty years of age with US $100,000 each to "focus on their work . . . research . . . and their self-education,"[30] with the caveat that they *not* attend university during the two-year fellowship. Both Chesky (in urbanism and community-building) and Thiel (in education and governance) imagine a better, more fulfilling world peopled by autopoetic micro-entrepreneurs—enterprising citizens free to take up and solve any challenge outside the constraints of race, gender, sexuality, class, and history. Such technologically mediated utopian desires remain stuck at the level of logistical form—centered on circulation, distribution, and consumption—rather than generating real change at the levels of production, labor, or value. They constitute attempts to rethink *process* without ever questioning the *system* in which those processes operate; and rather than imagining different futures, they remain trapped in a perpetual present, a cycle of unending creative destruction in which nothing fundamental can ever change. At the same time, with their calls to reshape how we imagine education, urban life, and government, they participate directly in expanding the field of enterprise far beyond Silicon Valley.

It is important to insist on the ways in which entrepreneurial subjectivity presents as potentially available to every segment of society and form of economic activity. In *Why Don't American Cities Burn*, Michael Katz describes in detail the shift of the language and social technologies used to name and address the problems of the crumbling cores of US cities and the poor who inhabit these spaces. From the 1970s to the present, the term *under-class*, "a concept covering up the old idea of the 'undeserving poor' with a veneer of social science," gave way to a celebration of "the entrepreneurial energy and talent latent within poor people who were waiting for a spark of opportunity to transform their lives."[31] An important step in this shift in the United States was the insistence in a series of articles by Harvard business professor Michael E. Porter (1995) about the entrepreneurial energies in inner-city communities—energies that simply needed the right kind of business-friendly programs to enable them to flourish, thereby producing the black and Hispanic middle-classes that generations of government programs had failed to create. For Porter, it was clear that "today's large and growing pool of talented minority managers represents a new generation of potential inner city entrepreneurs."[32] The expansion of programs designed to encourage poor entrepreneurs in the United States was mirrored by similar poverty-reduction strategies around the world, most notably by Nobel Prize winning economist Muhammed Yunus, who had helped found Grameen Bank, an organization designed to give loans to poor entrepreneurs.[33] In the words of C. K. Prahad, another business professor intent on saving the poor by including them in the world of profits, "if we stop thinking of the poor as victims or as a burden and start recognizing them as resilient and creative entrepreneurs and value-conscious consumers, a whole new world of opportunity will open up."[34] Of course, as Silvia Federici points out, an "ideology of micro-entrepreneurship" that sees the poor as creative entrepreneurs has the effect of "hid[ing] the work and exploitation involved" in financialized poverty alleviation, as well as the links between micro-entrepreneurship and debt rather than creation or job-satisfaction.[35]

The reassessment of those at the bottom of the social pyramid as entrepreneurs produced two important changes that have had an impact on the whole of society. First, it helped to legitimate the rollback or elimination of government social programs designed to help the poorest members of society, including direct transfers of resources via unemployment or welfare funds. The designation of these programs as "unsuccessful"—as unable to magically square the circle of capitalist accumulation and its generation of class divisions—paved the way for experiments in underclass entrepreneurship that continue to this day. Second, and perhaps even more significantly, the designation of entrepreneur as a new ideal of contemporary subjectivity has produced a change in how the poor understand themselves. Many now feel that they have no way to escape poverty other than by becoming entrepreneurs, especially given their extremely limited alternatives. Since the discourse of entrepreneurship insists that self-volition is all that is needed to generate results in our flattened post-society landscape, the outcome is that poverty can now only be a personal failing, not the consequence of social divisions, history, and the organization of power. The erasure of any gap between entrepreneurial success and personal effort and morality and the devolution of social, political, and economic issues onto atomized individuals have powerful effects; in the case of microfinance, they have instituted what Lamia Karim calls a "political economy of shame," wherein economic success or failure translates into a discourse of honor so that "honor and shame codes act as the collateral" for microloans.[36]

Although the development and promulgation of entrepreneurial subjectivity frequently appears, and thus often gets discussed as if it actually *is*, located in the Global North, as the structures of microfinance show, this model of subjectivity has spread around the globe.

Indeed, work on the global informal economy highlights the importance of an entrepreneurial spirit for making one's way as a member of
this 1.8-billion-strong workforce.[37] The ten-trillion-dollar-strong do-it-
yourself economy (*L'économie de la débrouillardise*), colloquially known
as "System D," demands entrepreneurial, innovative, and self-reliant
subjects;[38] the traits required for the kind of celebrated entrepreneurs
in Silicon Valley and Waterloo, Ontario, are just as necessary for the
hawkers, importers, bootleggers, market merchants, restaurateurs, scav-
engers, mechanics, and other entrepreneurial subjects whose labor takes
place off the books all over the world. Indeed, the prevalence of such
micro-entrepreneurs suggests that they, rather than the more large-
scale versions we generally point to (e.g., Bill Gates, Steve Jobs, Mark
Zuckerburg), might best demonstrate new entrepreneurial forms of
subjectivity. Further, if we think of the women accessing microfinance
and other *débrouillards* as performing entrepreneurship in the same
vein as that of the tech workers lauded in the pages of *Wired*, we may
begin to think that perhaps the common sense of entrepreneurial success that I have been describing through much of this essay, with its
belief in freedom to achieve on a level playing field that exists outside the
constraining barriers of privilege, in fact occupies the most privileged
position of all: once again we see that some playing fields are more level
than others.[39] Even so, the mode of entrepreneurship that is most commonly identified in the cultural imaginary with the first-world male tech
worker is a mode of subjectivity fast being adopted elsewhere. As the
New York Times notes, the IPO fantasies connected to the Chinese company Alibaba are an index of a broader social development in China.[40]
"The rise of Alibaba and its founder, Jack Ma, has proved instructive
for a generation of young Chinese—not just as a road map to riches,
but as a lesson in entrepreneurial individualism," write Andrew Jacobs
and Neil Gough, commenting that today, "thousands of young people
across China are creating start-ups of their own, driven by visions of
what they might do if they, too, strike it big."[41]

One last point needs to be made: not only are we all expected to be
entrepreneurs today, we are all expected to like it; from the perspective
of entrepreneurial common sense, *there are no unhappy entrepreneurs*.
Entrepreneurial subjects make no demands for any systemic change

on the system they inhabit. They inhabit it gladly, treating it as a space of unfettered possibility and excitement, and they tend not to fantasize different social futures, even if many entrepreneurs operate within a form of techno-utopianism in which they imagine that the combination of their activity and technological innovation will solve all the ills of both the individual subject and the planet as a whole. The dominance of entrepreneurial subjectivity therefore makes some recent assessments of affect in contemporary capitalism appear off the mark. Jane Elliott's description of personhood in neoliberalism as a form of "suffering agency,"[42] or Eva Illouz's discussion of the "cold intimacy" of capitalist life,[43] fail to capture the excitement and enticements of entrepreneurial being. In *The Structural Transformation of the Public Sphere*, Jürgen Habermas describes a process by which bourgeois society divides into the ferocity and coldness of the market and the warmth and sentimentality of family life, each pressured and undone by the necessity to maintain this artificial and ultimately unworkable division.[44] If this sharp division has broken down over time for a whole host of reasons (aesthetic and cultural as much as political and economic), the outcome is perhaps the opposite of what was imagined—not a society cold in all aspects of life, but one amenable to an active existence that no longer has time to occupy different affective categories at different moments.

Lauren Berlant's description of a "cruel optimism" that constitutes the dominant mode by which subjects experience contemporary society is a productive way of understanding the nature of entrepreneurial subjectivity.[45] But here, too, a key difference remains that makes entrepreneurialism at once a more productive and a milder mode of being. For Berlant, a tendency exists for contemporary subjects to "choose to ride the wave of the system of attachment that they are used to" instead of leaping into new, potentially better social forms.[46] This "reanchoring in the symptom's predictability" allows subjects to accept the limits of the quotidian and to project the capacity for change onto an unreachable horizon.[47] However, as the utopian situation for the entrepreneur remains always *the present*, cruel optimism transforms virtue into vice. The affect attendant to entrepreneurialism is not one that dissipates the energies for change through a faux reconciliation with the present, as mediated by optimistic fantasies of the future. Rather, it affirms the desirability of the present circumstances that enable entrepreneurialism and equates subjects' systems of attachment with an ideal system of belonging and behaving such that, even as entrepreneurs insist on the significance of their contributions in shaping the future, they occupy an ahistorical social landscape in which time stands still.

"The new subject is the person of competition and performance," write Dardot and Laval; "the self-entrepreneur is a being made to 'succeed,' to 'win'." The "performance/pleasure" apparatus that Dardot and Laval outline captures the dynamic at the heart of contemporary entrepreneurship.[48] The subject of contemporary capitalism no longer participates in a producing/saving/consuming cycle that balances the sacrifice of work with the goods and pleasures attained as its result. Instead, today's subject participates in an economy of surplus. The demand to *produce* ever more is part of a system in which an imperative exists to *enjoy* and to *become* ever more. The injunction for maximum performance in life *and* work, exemplified for Dardot and Laval in professional sports, has become mandatory throughout the social field: "subjects are enjoined to 'surpass themselves,' to 'push back the limits' as managers and trainers say."[49] The imperative for surplus constitutive of the entrepreneurial subject fundamentally alters the power dynamic in the work relationship. The management of subjectivity—its array of desires, beliefs, and behaviors—for optimal work performance, which has been at the heart of the discipline of business administration (and even more so for human resources), has become more and more important as profits have declined and affective forms of labor have become a more prominent feature of Western economies. Entrepreneurial subjectivity requires no external monitoring, no system or apparatus by means of which the individual is adapted to the needs and structures of the enterprise. Instead, the subject of production and the psychological subject are one and the same: they are always already identical, insofar as (for the entrepreneur) creating an enterprise and creating a self *is the same activity*. The endless drive to exceed one's capacities across hitherto distinct spheres of life activity—at work, at the gym, as a parent, as an investor—creates a subjectivity that produces as much as possible in as many areas as possible, and it does so at every waking moment.

Importantly, the imperative for pleasure and performance attaches not only to success but also (and perhaps more securely) to *failure*. Entrepreneurs have unrealistic ideas of success and unhealthy fantasies about the productivity and necessity of failure. A study of 3000 entrepreneurs revealed that 81 percent believed that their businesses had 70 percent or greater chance of success—a wild overestimate. In addition, entrepreneurs who have failed previously are likely to fail again: a 2009 study revealed that first-time failures would come to the same end, and, indeed, that second-time entrepreneurs would fail even more often.[50] The likelihood of failures is not buried under the ideology of success—the operant ideology of liberal capitalism in which each and every person can become middle-class bourgeois subjects. On the contrary: entrepreneurial subjects embrace and even seek out failure as an important, even *essential*, dimension of their activity. Samuel Beckett's induction in *Worstward Ho* to "Try again. Fail again. Fail better"[51] has been taken up as a defining motto of entrepreneurial culture, which thrives on the idea of risk, and which celebrates the extreme risks that might produce short-term failures potentially generating long opportunities and openings in the market never before imagined.[52] For the entrepreneurial subject, failure might well be more important than success. The cyclical process of risk and failure enacts a classical version of Sigmund Freud's repetition compulsion: if you fail, start something new; if you succeed, launch an initial public offering, get the hell out . . . and start something new. And yet, just as with cruel optimism, the compulsion to enact failure produces a different outcome than that imagined by Freud: neither belated mastery of a traumatic event nor a form of repression in need of an outlet, the entrepreneurial repetition instead bolsters capacity and agency. The repetition of failure becomes a badge of pride, a marker of living well, of engaging in properly ethical behavior, and of having achieved the good life.

If one had to describe the vision of the world outlined by neoliberalism in a sentence, it might be this: it is now the market that supplies the state with its principles and mandate, rather than the state guiding, shaping, and supervising the market on behalf of those subjects who (at least in theory) collectively legitimate the state's actions and practices.

The multiple claims and demands that the logic of entrepreneurship makes on us constitute an incredibly powerful and appealing common sense, one increasingly difficult to challenge or critique; in a world of risk and uncertainty, this logic not only offers possibilities of order and control but also makes claims on human flourishing and self-fashioning. We have now all been given the freedom to become a new kind of petite bourgeoisie, but without the constraints of bourgeois morality or the crippling desire to become anything other than ourselves; and in the process of our work as entrepreneurs, we will once and for all get rid of the limits of wage labor and its far-too-crude (and now antiquated) modes of generating value. These are class fantasies shaped to manage and disavow the very real collective challenges that we face—and not just with respect to socio-economic justice: no amount of social entrepreneurship will be able to address the ills of global capitalism, no matter how many apps and websites pop up aiming to address poverty or the maldistribution of income.

The rapid rise of the entrepreneurial subject indexes a change in the structure and organization of power with which we have yet to fully grapple. The status of entrepreneurship as a new common sense of subjectivity and economic practice—an accepted mode through which a congruity is achieved between a responsible, moral, self-fashioning individual and an economic-rational individual—would suggest that it constitutes an ideal subjectivity for neoliberal forms of governmentality, one that it has been searching for all along. Thomas Lemke reminds us that "neo-liberal forms of government feature not only direct intervention by means of empowered and specialized state apparatuses, but also characteristically develop indirect techniques for leading and controlling individuals without at the same time being responsible for them."[53] Entrepreneurship would appear at first glance to exemplify such a mode of indirect control sans responsibility. Yet as much as it might thus represent the smooth operations of contemporary governmentality, it may be that the fast embrace of its logics also poses a challenge and a threat to state apparatuses. While it may begin as an indirect technique of control, the discourse of entrepreneurship makes it clear that this practice lies *outside* of and beyond the claims of government. It is a mechanism of selfhood and subject formation that begins from the premise that there is no one to count on, no one who can do anything for you other than you yourself.

The production of subjects responsible only for themselves—a responsibility measured primarily via the dictates of the market—means that, from the perspective of such subjects, even a stripped-down state apparatus is one that does not properly function in the twenty-first century. Today, the state is outmoded, for whatever else the state can be it cannot be entrepreneurial by definition. The state has had to insist on the production of entrepreneurial subjects because it itself is unable to respond to the demands and insistences of a marketplace fueled by a desperate search for value anywhere and everywhere. Governments cannot be entrepreneurial, nor can NGOs. However, luckily, entrepreneurs can use their unique insights and ideas to address the built-in limits of governments, a barrier that is one of scale but also of temporality. In generating subjects to address the risks faced by the state—especially risks connected to the production of value, as Melanie Gilligan and Marina Vishmidt point out—the state hopes to shirk responsibility while netting the economic benefit of entrepreneurs.[54] However, why entrepreneurial subjects would see government as anything other than an inefficient, expensive apparatus in an era when most apps are free and can be downloaded on one's phone with the touch of a smartphone screen is open to question. Entrepreneurship may be simultaneously the height of neoliberal subject formation *and* its limit—a peak on the other side of which lie subjects with no fidelity or loyalty to governments or states.

We are far from the end or even significant reconstitution of the biopolitical state apparatuses that Michel Foucault analyzed so productively in his late lectures. And the downloading of responsibility onto subjects in an era in which that responsibility is configured primarily in relation to participation in the market hardly constitutes the end of the state or the creation of new subjects and communities in relation to it. And yet, should we not welcome the cracks that might appear in the operations of biopolitics at its fullest operation? Even as we might be cautious, wary, and worried about the creation of entrepreneurial subjects—for all the reasons we outline here—might there also be, strangely, openings in its rejection of a neoliberal state apparatus that has otherwise appeared to operate as if nothing could stand in its way? The mania for entrepreneurship that exists today in all manner of social, cultural, and economic practices point to a terrifyingly complete accommodation and identification of the self with the market, a brave new world made up of subjects who believe they exist in a society freed of the castes that Aldous Huxley imagined would comprise a future command economy. At a moment in which the state of power can be captured instantly by invocations of the 1 percent versus the 99 percent, this cannot help but be alarming. Still, there is a kernel of political possibility, a hint of imaginative self-reliance and rejection of the status quo, in the desire to produce one's own life, failure or no, against the dictates of class or origin, that speaks to political inventiveness and possibilities just over the horizon.

NOTES

1. Gideon Lewis-Kraus, "One Startup's Struggle to Survive the Silicon Valley Gold Rush," *Wired*, Apr. 22, 2014, 110, www.wired.com/2014/04/no-exit/ (accessed August 17, 2018).

2. Lewis-Kraus, "One Startup's Struggle," 110.

3. See, e.g., Tardy, *The Eventual Millionaire*; Sisson, *The Suitcase Entrepreneur*; Brogan, *The Freaks Shall Inherit the Earth*.

4. After beginning in Japan as *The Tigers of Money*, *Dragons' Den* has gone on to have nation-specific versions in twenty-six countries. Along with *Shark Tank*, the American version of *Dragons' Den*, viewers in the United States have been treated to *American Inventor* and *The Profit*, British viewers can enjoy *The Big Idea*, and Chinese viewers can check out entrepreneurial wannabes on *Yíng Zài Zhōngguó* or *Win in China*.

5. As part of its marketing strategy, Mattel has linked Entrepreneur Barbie to eight female entrepreneurs, who act as her "Chief Inspirational Officers." They include Reshma Saujani, founder of Girls Who Code, Deborah Jackson, founder of microfunding site Plum Alley, and Gina Rudan, founder of leadership training company Genuine Insights. See Laura Hudson, "Your First Look at Entrepreneur Barbie, Smartphone and All," *Wired On-line*, June 18, 2014, http://www.wired .com/2014/06/entrepreneur-barbie/ (accessed August 17, 2018).

6. Business incubators and accelerators are intended to speed up the development of business ideas, to weed out those deemed untenable, and to link entrepreneurs with investors and potential clients; while some provide funding, many only offer guidance. Although total numbers of such organizations are not available, the US–based National Business Incubation Association (2014) lists 2200 members from 62 countries.

7. A few Canadian examples include the Entrepreneurship HUB at the University of Alberta (http://entrepreneurship.ualberta.ca/ehub); the Wilson Centre for Entrepreneurial Excellence at the University of Saskatchewan (http://www.usask

.ca/wilsoncentre); and the University of Toronto's Entrepreneurship Hatchery (http://hatchery.engineering.utoronto.ca). US examples include the National Collegiate Inventors and Innovators Alliance, which partners with 200 colleges and universities to support student and faculty entrepreneurship (http://nciia.org); the University of Wisconsin-Milwaukee's Start-up Challenge (https://uwm.edu/lubar-entrepreneurship -center/student-startup-challenge/); and the Arthur M. Blank Center at Babson College (http://www.babson.edu/Academics/centers/blank-center).

8. Toronto alone houses two centers for entrepreneurial artists: Artscape, which offers a "Creative Entrepreneurship Program" (http://www.artscape.ca), and the Toronto Centre for Entrepreneurial Art and Design (http://www.tcead.com/). For an examination of the relationship between artistic practice and entrepreneurship, see Vishmidt, "Mimesis of the Hardened and Alienated," and Shukaitis and Figiel, "The Factory of Individuation."

9. Cantillon, *An Essay on Economic Theory.*

10. Whyte, *The Organization Man.*

11. Hall and O'Shea, "Common-sense Neoliberalism," 8.

12. Vital, "Everyone Will Have to Become an Entrepreneur."

13. Graeber, "On the Phenomenon of Bullshit Jobs."

14. See also Kolbert, "No Time."

15. To give one example, in 2014, the Government of Ontario pledged $51 million CAD to a range of programs that "provide young entrepreneurs with business skills, capital to start and grow a small business and the opportunity to gain experience and expertise through R&D internships." Ontario Ministry of Research and Innovation, "News Release: Helping Students Start Innovative Companies, New Youth Jobs Programs Fostering Entrepreneurs," *News room*, Apr. 24, 2014, www.news.ontario.ca/mri /en/2014/04/helping-students-start-innovative-companies.html (accessed August 17, 2018).

16. Council of Ontario Universities, *Entrepreneurship at Ontario Universities*, October 2013, 1, http://cou.on.ca/reports/entrepreneurship/ (accessed August 17, 2018).

17. The scope of programs in entrepreneurial studies makes even a partial survey difficult. To offer just a small sense of the embeddedness of entrepreneurship as a field of study in the contemporary university, consider the now numerous centers and institutes devoted to the subject across the United States, including the Greif Center for Entrepreneurial Studies at the University of Southern California, the Wolff Center for Entrepreneurship at University of Houston, and the Center for Entrepreneurial Studies at Stanford University. State colleges are equally well endowed with entrepreneurship centers: there are top-ranked programs of study offered by the University of Missouri-Kansas City's Regnier Institute for Entrepreneurship and Innovation and the University of Oklahoma's Center for Entrepreneurship. In 2011–12, the program offered by the University of Houston had 2000 registered students.

18. See, for instance, "Venture Capital Monthly," a report produced by the company Pitchbook, which ranks universities according to their ability to produce venture capital–backed entrepreneurs. By this measure, the fourth best university in the world is the Indian Institute of Technology; Cambridge, Oxford, and the University of London do not place in the top fifty.

19. NMU College of business, "Entrepreneurship Major," *NMU College of Business*, 2014, www.nmu.edu/business/node/59 (accessed August 17, 2018).

20. This is in essence Joseph Schumpeter's oft-cited view of the need for entrepreneurs in capitalism: "the carrying out of new combinations we call 'enterprise'; the individuals whose function it is to carry them out we call 'entrepreneurs.'" Schumpeter identified five new "combinations" that entrepreneurs can undertake: new goods, new methods of production, the opening up of new markets, the creation of a capacity to use new raw materials, and reorganization of an industry (i.e., the breaking up of a monopoly situation). Schumpeter, *The Theory of Economic Development*, 65–66.

21. For an account of the crisis situation of contemporary capital, which includes a long-term decline in the rate of economic growth, see Streeck, "How Will Capitalism End?"

22. In addition to Beck, *Risk Society*; Butler, *Precarious Life*; and Agamben, *Homo Sacer*; see, among others, Southwood, *Non-Stop Inertia*; Fisher, *Capitalist Realism*; Davis, *Planet of Slums*; Bauman, *Wasted Lives*; and Harvey, *The New Imperialism*.

23. Rawls, "Justice as Fairness."

24. Dardot and Laval, "The New Way of the World, Part II."

25. Slavoj Žižek offers a helpful reminder of the distinction between formal and actual freedom: "Formal freedom is the freedom of choice within the coordinates of the existing power relations, while actual freedom designates the site of an intervention that undermines these very coordinates." Žižek, *On Belief*, 122.

26. See www.seasteading.org and www.thielfellowship.org, respectively. For more on Thiel's ideas on education and Thiel Fellowships, see Shaffer, "Back to the Future with Peter Thiel" *National Review Online*, Jan. 2011, www.nationalreview.com /articles/257531/back-future-peter-thiel-interview (accessed August 17, 2018); Caitlin Kelly, "The Theil Fellows, Forgoing College to Pursue Dreams," *New York Times*, Sept. 15, 2012; and Alison Griswold, "The Dropout Fallacy," *Slate*, May 23, 2014, http://www.slate.com/articles/business/how_failure_breeds_success/2014/05 /peter_thiel_drop out_grant_encouraging_students_to_stop_out_of_college.html (accessed August 17, 2018). For more on the seasteading movement, see China Mieville, "Floating Utopias," *In These Times*, Sept. 28, 2007, www.inthesetimes.com /article/3328/floating_utopias (accessed August 17, 2018).

27. For more on the sharing economy, see Gansky, *The Mesh*; "All Eyes on the Sharing Economy," *The Economist*, Mar. 9, 2013; Nanos, "The End of Ownership," *Boston Magazine*, May 2013; and Thomas Friedman, "Welcome to the Sharing Economy," *New York Times*, July 20, 2013.

28. Brian Chesky, "Shared City."

29. Seasteading Institute, "FAQ: Vision, Mission, and Goals," seasteadinginstitute.org, 2012. www.sea-steading.org/about/faq/.

30. See www.thielfoundation.org (accessed August 17, 2018).

31. Katz, *Why Don't American Cities Burn?*, 101. For additional details, see especially chapter 4, "From Underclass to Entrepreneur," 101–50.

32. Porter, "The Competitive Advantage of the Inner City," 62.

33. See Yunus, *Banker to the Poor*. For critiques of microfinance as a means of alleviating poverty and empowering women, see Karim, *Microfinance and Its Discontents*;

Vishmidt, "Permanent Reproductive Crisis"; and Federici, "From Commoning to Debt."

34. Prahad, *The Fortune at the Bottom of the Pyramid*, 1.

35. Federici, "From Commoning to Debt," 239.

36. Karim, "Demystifying Micro-credit," 10.

37. Williams and Nadin, "Entrepreneurship and the Informal Economy," 364.

38. Neuwirth, *Stealth of Nations*. For more on the informal economy, see Davis, *Planet of Slums*; Venkatesh, *Off the Books*; Williams and Nadin, "Entrepreneurship and the Informal Economy"; and Harris, *The International Bank of Bob*.

39. For an excellent overview of the forms of exploitation and violence effaced in the common sense of entrepreneurial possibility, see Mies, *Patriarchy and Accumulation on a World Scale*.

40. Andrew Jacobs and Neil Gough, "Alibaba, With Its I.P.O., Mints Millionaires and Risk-Takers," *New York Times*, Sep. 18, 2014.

41. Jacobs and Gough, "Alibaba."

42. Elliott, "Suffering Agency."

43. Illouz, *Cold Intimacies*.

44. Habermas, *The Structural Transformation of the Public Sphere*.

45. Berlant, "Cruel Optimism."

46. Berlant, "Cruel Optimism," 23.

47. Berlant, "Cruel Optimism," 31.

48. Dardot and Laval, "The New Way of the World, Part II."

49. Dardot and Laval, "The New Way of the World, Part II."

50. See James Surowiecki, "Epic Fails of the Startup World," *New Yorker*, May 19, 2014, 36, www.newyorker.com/magazine/2014/05/19/epic fails of the startup-world (accessed August 17, 2018).

51. Beckett, *Company; Ill Seen Ill Said; Worstward Ho; Stirrings Still*, 81.

52. Beckett's phrase from *Worstward Ho* reads, "Ever tried. Ever failed. No matter. Try again. Fail again. Fail better." Beckett, 81. *Slate* columnist Mark O'Connell claims, "The entrepreneurial class has adopted the phrase with particular enthusiasm, as a battle cry for a startup culture in which failure has come to be fetishized, even valorized." O'Connell, "The Stunning Process of 'Fail Better': How Samuel Beckett Became Silicon Valley's Life Coach," *Slate*, Jan. 29, 2014, http://www.slate.com/articles/arts/culturebox/2014/01/samuel_beckett_s_quote_fail_better_becomes_the_mantra_of_silicon_valley.html (accessed August 17, 2018).

53. Lemke, "'The Birth of Bio-Politics,'" 201.

54. Gilligan and Vishmidt, "'The Property-less Sensorium.'"

CHAPTER 10

Conjectures on World Energy Literature (2017)

Energy is a fundamental element of human life.[1] It is what gives societies the capacity to undertake all of the varied activities that occur within them—what, in a word, makes societies and the individuals in them "go," and not only in a basic, material sense, but socially and culturally, too. To take but one example: energy is required for the modern system of transportation to function; what makes that system "go" is not only the raw stuff of fossil fuels (and, to a much lesser degree, electricity), but also the ideas and ideals driving everything from global trade to individual mobility—the "freedoms" associated with neoliberal economics and the ability to travel when and where one wants.[2]

If the first point might seem obvious—human societies require energy if they are to function at all—the latter point—that there is a connection between energy and culture—has only recently come to the fore of critical attention.[3] For too long, energy has been treated as a largely neutral input into societies—a necessary element of social life, but not one that has any significant, defining impact on its shape, form, and character. The history of modernity, for instance, has been figured in relation to novel developments in literary culture, scientific discovery, the birth of cities, the expansion of individual political freedoms, the structures and strictures of colonialism, and the creation of the global nation-state system. It has seldom been narrated in relation to the massive expansion of socially available energy (and energy of a specific form: fossil fuels, which are easy to transport and store) and the concurrent redefinition of social practices, behaviors, and beliefs occasioned by this historically unprecedented explosion of access to energy.[4] Leaving energy out of our picture of modernity (and indeed, out of any period of history) has meant that we lack a full understanding of the forces and practices animating not only its politics and economics, but also its social and cultural life. Energy names the material force that makes societies go. But energy is also a social relation, animating through its specific form the character and capacities of the societies we inhabit and the practices that shape them. As we attempt to address the consequences of modern petrocultures and to transition beyond them, we need to include energy in our account of any and every aspect of the social and cultural landscapes we inhabit. Without doing so, we remain stuck, imagining techno-utopian solutions to our environmental crisis instead of getting to work re-defining our existing social relations, structures, and behaviors.[5]

What does this mean for the study of literature and of culture more generally? Environmental scientist Vaclav Smil, who has repeatedly drawn attention to the broad historical and social significance of energy, has written: "timeless artistic expressions show no correlation with levels or kinds of energy consumption: the bison in the famous cave paintings of Altamira are not less elegant than Picasso's bulls drawn nearly 15,000 years later."[6] This view of the relationship between energy and cultural production is sustainable only if one imagines the aesthetic as falling completely outside history—as in fact "timeless" in the way Smil suggests—and is in fact contradicted by Smil's own comments elsewhere in "World History and Energy" about the depth of energy's impact on the development of the forms taken by societies throughout history.[7] Smil's claims about the lack of a relationship between art and energy come in the final section of his essay, in which he cautions against a simplistic "energy determinism" in relation to historical development. One of the significant challenges of introducing energy into cultural and social analysis is to figure out exactly how to plot its impacts and effects. While adding energy to the study of literature and culture might not constitute something akin to a "final" explanation of these practices and their symbolic significance (as if energy were a substitute for the economic base, "in the last instance" determining *both* the character of the economic base and the superstructure connected to it), it is equally problematic to write energy out of the picture, as having no significance at all, and so requiring none of our critical and conceptual attention when we engage in the study of literature and culture.

An energy determinism is unsustainable. The claim that energy has no impact on literature and culture is equally unsustainable, whether we understand this impact in a narrowly material sense—in the very substance of the acrylic paints used on modern canvases, the stock used to shoot films, or in the electricity required to run printing presses and generate cable signals—or socially, through its figuration of social capacities and expectations.[8] How might one begin to add energy to cultural analysis in a way that captures its force and impact, without deferring to either one of these extremes? This is the question that I want to take up, specifically in relation to how energy might add to our accounts and understanding of world literature. Does making a link between a specific energy system and a literary period (or movement, or form) open up a new way of analyzing texts? Or does it unnerve not just the how but the *why* of literary studies? Given the significance of energy for societies, we need to begin to add energy to our literary and cultural analysis. But just what does it mean to do so, especially in our accounts and understanding of world literature?

* * *

The history of energy is constituted by forms of energy giving way to other forms of energy—for instance, the dominance of water and wind in the United Kingdom in the early nineteenth century moving to coal by the mid to late part of the century.[9] One of the ways in which it might thus be possible to grasp the impact of energy on literary and cultural production is by engaging in a process of periodization—refiguring literary history around eras defined by wood, tallow, coal, whale oil, gasoline, and atomic power, as in the title of Patricia Yaeger's provocative *PMLA* editor's column on literature and energy.[10]

The aim and intention of Yaeger's loose periodization of energy is to begin a conversation about the material and social impact of energy within literary studies. Such broad energocultural periodizations are rendered immediately problematic by the real character of energy systems. In our own energy period, it can be easy to overstate the degree to which oil or fossil fuels are in fact the dominant form of energy.[11] To begin with, the forms of energy used at any given moment in history are inevitably mixed and changing. While the use of fossil fuels since 1850 has decisively re-shaped the sources of energy used,[12] the specific mix of the types of fossil fuels used (coal, gas, and oil) shifts and changes year in and year out. And as new forms of energy are discovered and added to the mix, energy systems become more and more diverse and complicated; contemporary energy systems are made up of an array of energy forms—hydro, nuclear, solar, wind, and more in addition to fossil fuels (though new forms have yet to have a major impact: for instance, the introduction of nuclear energy after World War II has done very little to offset growth in the use of fossil fuels). In an era that I've come to describe as a "petroculture" to try to stress the ultimate importance of oil, the use of coal continues to grow, and not only in the developing world.[13]

What might further trouble any easy definition of energy periods are the vast differences in levels and forms of energy use around the world. This is true both between developed nations, and even more so between developing and developed nations. Canada and Germany are both members of the G8 and are among the group of the world's most developed nations. However, the mix of energy forms that they use and the level of energy employed per capita makes it difficult to easily figure them as part of a single petroculture: Germany has a far greater proportion of sustainable energy than Canada as part of its overall energy use, and uses about half of the energy per capita as Canada.[14] When the comparison is made between developed and developing countries, the direct link between access to energy and levels of development becomes all too clear. Per capita energy use in Canada is *twenty-five* times greater than in the Democratic Republic of Congo, and close to *twenty* times that in Benin and Haiti. These differences in energy use are connected to histories of colonialism, underdevelopment, and global political and economic power; even in petromodernity, large swathes of the planet are still powered by animal labor and the labor of bodies. Finally, the assertion of a single petroculture obscures the huge differences of access to power *within* nations. Elites everywhere use more energy per capita than poorer members of societies. While some elites have no doubt transitioned from fossil fuels to solar power on their homes, the use of animal energy, wood, and coal continues to fuel societies around the world.[15] Does this admixture of energy forms render the task of energy periodization confusing and pointless, unable to provide analytic insight into the impact of energy on culture?

Despite the complex map of energy use at the present time, it seems to me that energy periodization *can* open up new ways of figuring and analyzing literature in relation to the world in which it is produced. To begin with, where and what kinds of energy are used, and how much is used and by whom, provides a map of power, privilege, control, and dispossession. There is a strong correlation between energy use and GDP; increases in energy use per capita inevitably lead to economic growth and higher living standards. The politics of colonialism and postcolonialism were shaped by race and ethnicity, military power, and control over the trade of goods (with colonial powers benefiting at the expense of the colonies they controlled). Colonialism was also a period animated by access to energy, and the search for and the struggle to control ever more energy. Access to fossil-fuel energy—first coal, and then oil and gas—was essential to managing and maintaining colonial control; the struggle over energy remains a largely untold story within postcolonial literary studies and in the study of world literature, which has focused on other aspects of the combined and uneven development that has characterized modernity.[16] At a minimum, tracking levels of energy use can be used to identify nodes of power in the world, and to consider the impact of these differentials of power in relation to cultural production. Indeed, one could argue that the "world" announced by the category of "world literature" doesn't in fact come into existence until the beginning of the era of fossil fuels: the production of the imaginary named "world" is fueled by the presence of energy sources, including coal and oil, that make the space of the globe increasingly available and accessible to travel, trade, and political power.[17] A global map of energy, one that also captures differences of energy use within nations, highlights the ways in which power is actively materialized, with inevitable impacts on social and cultural capacity and sensibility.

With few exceptions—Saudi Arabia and the Emirates stand out today as not just producers of energy but also as enormous energy consumers—a map of energy power is, however, not unlike our already existing understanding of power and privilege, from colonialism through to the postcolonial, neoliberal present. The strong connection between energy and political power, between levels of per capita energy use and socio-economic status, means that existing narratives of global power tell us a great deal about the socio-cultural impact of energy use, even if the specific importance of energy is rarely named. Can we thus do without critical appeals to energy, assuming that its force and social import will show up in the political, economic, and military structures and forces that it has fueled?

It seems clear to me that this would be a mistake. While it may be the case that there is a strong correlation between existing narratives of culture and power, the specific role of energy in shaping both cannot just be passed over, nor can the specificity of the experiences, social sensibilities, and cultural imaginaries produced by distinct energy systems. There is also a strong political rationale for foregrounding fossil fuels in the analysis of cultural production. Given the strong link between expanded energy use and development, countries in the Global South will of necessity need to use more and more energy per capita in order to give their citizens the social capacities and opportunities that high levels of energy use have afforded in the Global North. Whatever forms development might take, it means using more and more energy.[18] The environmental consequences of doing so, and the social and political pressures this will generate at a global level, promise to extend existing colonial and postcolonial struggles deep into the twenty-first century, and do so in reference to an issue to which we barely paid any attention—energy.

Despite the mixed and uneven use of energy around the world at the present time and throughout modernity, it *is* possible to speak meaningfully of a period of petromodernity. Petroleum is the *hegemonic* form of energy at the present time. The claim that something has become hegemonic is intended to capture an organizing principle or a shaping dictate of a period: what is hegemonic about immaterial labor at the present time, for example, is not that the majority of people are involved in such forms of work, but that it places demands on all forms of labor to "informationalize, become intelligent, become communicative, become affective."[19] It is in this sense that fossil fuels are hegemonic and that we live in a petroculture. Energy use remains mixed, both in terms of the forms of energy use and the amounts used by individuals around the world. However, our cities have not been shaped around nuclear energy or hydropower. The fantasy of suburban living and the freedoms of highways owe nothing to wind farms and solar power; and no country imagines that the way forward in their development is to shape agricultural systems around plow and oxen as opposed to mechanical farming and fertilizers. The enormous, unprecedented energy of fossil fuels has shaped (and continues to shape) our cultural and social imaginary in profound ways, including the character of our political structures and, as Timothy Mitchell points out, the principles and rationales around which we organize our economic practices and decisions.[20] Being alert to the differences and distinctions in energy use that I've already pointed to is essential. My intent in claiming that we understand modernity as a petromodernity isn't to flatten out these differences, in a manner akin to the way that the Anthropocene has been critiqued for eliminating a more nuanced history of the human impact on the environment.[21] It is, rather, to insist that we also attend to the force and power of the dominant rationales and logics that have shaped the world in which these differences and distinctions exist, including those connected to the forms of energy in use.

In "Conjectures on World Literature," Franco Moretti argues that we need to understand the world literary system as "one, and unequal: *one* literature (*Weltliteratur*, singular, as in Goethe and Marx) . . . but a system which is different from what Goethe and Marx had hoped for, because it is profoundly unequal."[22] So, too, with the system of energy use: one petroculture, but a profoundly unequal one. Moretti's essay provides a quick articulation of and justification for his system of "distant reading," which is intended in part to identify dominant logics that have structured the world literary system. For him, the system of *Weltliteratur* is organized by a triangle of forces: "foreign form, local material—and *local form*."[23] Moretti cites the Brazilian literary critic Roberto Schwarz, who provides an example of what the inequality of world literature produces at the level of content: "Foreign debt is as inevitable in Brazilian letters as in any other field . . . it is not simply an easily dispensable part of the work in which it appears, but a complex feature of it."[24] One might imagine inequalities not only of financial power (which generates debt) but also of energy form and capacity to be figured differently at center and periphery, shaping and forming what Moretti calls "local materials" in a strong and profound way. Jennifer Wenzel's "Petro-Magic-Realism" offers an example of literary analysis that foregrounds the impact of energy on literary form at the periphery. In order to offer further specificity and analytic density to the widely used literary category of "magic realism," Wenzel argues that the "political ecology" of Nigerian literature is better described as "petro-magic-realism," "a literary mode that combines the transmogrifying creatures and liminal space of the forest in Yoruba narrative tradition with the monstrous-but-mundane violence of oil exploration and extraction, the state violence that supports it, and the environmental degradation that it causes."[25]

Such mappings of specific encounters between aesthetics and politics via the position of literature within petromodernity are essential; there are all too many oil encounters about which the whole story has yet to be told. If I had to assign a tendency that has emerged in literary critical attention to energy systems, and especially to the petrocultural system that has shaped the colonial encounter and the experience of the modern, it is the attempt to identify the sites and spaces in which the specific configuration of form and content names not only a political tension but also a tension animated by energy differentials (unlike the Marxist critics such as Masao Miyoshi, Fredric Jameson, and Nicholas Brown whose attention stays more squarely in the realm of political economy). But the inclusion of energy inequality as an axis of analysis is hindered by one major problem—a defining aspect of oil in relation to culture that has to be placed at the center of any conceptualization of energy as an analytic tool within the study of world literature. *The importance of fossil fuels in defining modernity has stood in inverse relationship to their presence in our cultural and social imaginaries,* a fact that comes as a revelatory surprise to almost everyone who engages in critical explorations of energy today. The arts of a world powered by horses and the labor of bodies cannot help but be distinct from the expanded time, space, and power of our own petromodernity. Nevertheless, attention to energy differentials has largely been absent not only from critical investigations of literature and culture, but also from literature and culture itself. Those who have begun to grapple with the cultural absence of oil in a period shaped by and around the substance have scanned literary history to find texts that have confronted oil hegemony, with productive results.[26] But questions remain. Why has it been so difficult to locate a literary archive for what now appears to be so important a social and historical force? And despite this absence or gap, why does it seem necessary and important to do so now?

One of the earliest links between energy and aesthetics was made by Amitav Ghosh in his 1992 essay "Petrofiction: The Oil Encounter and the Novel." Ghosh's essay is, in part, a review of Abdul Rahman Munif's monumental *Cities of Salt* quintet, which maps the coming-to-be of American petroleum interests in the Middle East and its world-altering impact on those living in the Arabian Peninsula. "Petrofiction" also offers a broader critique of the failure of literary fiction to address what Ghosh names the "Oil Encounter": the intertwining of the fates of Americans and those living in the Middle East around this commodity, a connection that continues to have far-reaching economic, cultural, and political reverberations.

In the twentieth century, US power was strongly tied to oil. For the first half of the century, the United States was the world's first true oil superpower, and it used its energy riches to develop a consumer culture built around automobiles, suburbs, and malls. The geopolitical crises of the second half of the last century and the first two decades of the current one are the result of the need for the United States to continue to have access to oil to fill the tanks of SUVs as well as fuel its larger economy. Given the deep imbrication of the US economy and its politics with petroleum, Ghosh expresses puzzlement that no author has taken up the Oil Encounter as literary subject. Ghosh's essay is now a quarter-of-a-century old; and yet there are still far too few literary fictions that undertake an exploration of what could well be said to have defined the politics of the period since the discovery of oil in the mid-nineteenth century—the struggle over access to oil. Nor, for that matter, are there very many fictions that have dealt with the social, cultural, and political importance of energy and oil in any way, shape, or form. I'm not suggesting that there are no such fictions—there are, including such one-of-a-kind theory-fictions like Reza Negarastani's *Cyclonopedia: Complicity with Anonymous Materials* (2008) and recent postcolonial narratives such as Helon Habila's *Oil on Water* (2011).[27] Very often, however, the presence of oil in fiction is little more than shorthand for wealth and power, which could have other sources and be just as well figured in other ways.

Ghosh's insight regarding the absence of oil and the Middle East from modern US fiction has resonated with critics and theorists who have attended to the representational challenge presented by fossil fuels. As the planet's hegemon during the height of petromodernity, we might well expect to find it difficult for US fiction and culture to figure, directly and unfalteringly, one of the principal sites and sources of the nation's power. From the perspective of the United States, oil is simply what makes the country "go," and in a way that doesn't necessitate comment or concern; as a result, it gets lost to the background of the physical apparatus within and against which social, cultural, and political life is played out, no more worthy of comment than furniture or the asphalt covering its streets—indeed, less so.

One of the distinctive elements of oil as an energy source is that, unlike coal-energy (which does enter into Victorian-era writing; see MacDuffie's *Victorian Literature, Energy, and the Ecological Imagination*), it is a resource whose consumption is disassociated from its extraction. The historian Christopher Jones points out that from the very beginning of the establishment of the oil pipeline system, "the users of oil gained the benefit of cheap energy without assuming responsibility for its environmental damage."[28] This ethical dissociation also speaks to its place in the cultural and social imaginary—it vanishes to the background, invisible to narrative and so, too, to critique. What Munif's *Cities of Salt* (1987) captures so perfectly is the blind sense of entitlement by those engaged in extracting oil, and a sense of puzzlement by those native to the land whose space and culture have been invaded. Even more powerfully, it shows how the world made by oil can only be seen by those foreign to it. *Cities of Salt* manages to transform oil modernity into the science fictional time and space that it is: occupied by creatures of steel and asphalt that are animated by the liquid remains of plants millions of years old, and whose imperatives and rationales seem so out of joint with the physical environment of the Earth that it is hard to believe that they are not from elsewhere in the universe. "What new era had begun—what could they expect of the future? For how long could the men stand it? The night had passed, but what about the nights to come?"[29] The fear and uncertainty that Ibn Rashed and his community experience in *Cities of Salt* is justified. The dissociation that enlivens the Americans is infectious; their appearance in the Middle East to hunt for the substance that fuels them is felt by the Arab community as a rupture in time with unknown consequences.

Munif and other postcolonial writers have managed to grasp some of the social power of oil as a result of imaginative openings that the periphery provides. Just as the colonial system and the mechanisms of postcolonial power are evident to those who have to endure them, one might imagine that the significance of energy for shaping modernity is readily apparent in those spaces where petroculture has yet to take hold. Munif's *Cities of Salt* shows the critical openings that the gap between center and periphery can provide. And yet, the representational challenges posed by oil are deeper and more complex than this. Ghosh's interest in Munif's work arises in part because it stands as an exception: it is not only US fiction that fails to name oil as a key element of the human drama of the twentieth century, but much of postcolonial fiction as well. For the most part, despite their evident socio-historical importance, fossil fuels have in fact played a very small role in artistic and literary expression during the fossil-fuel era. In this case, the exceptions—those texts that have been examined by literary critics working in the energy humanities—definitely prove the rule.

I suggested earlier that whether or not it was used everywhere in the world to the same degree (or indeed, whether it was used at all in some places), oil has to be seen as hegemonic—as an energy source that organizes life practice in a more fundamental way than we've ever allowed ourselves to believe. What this means is that, despite the absence of fictions that take up oil or the oil encounter directly by making it a part of narrative, form, or both, *any and all* examples of cultural expression in the era of oil have in fact been crucially figured by it.[30] In their introduction to *Oil Culture*, Ross Barrett and Daniel Worden write that "oil culture encompasses the fundamental semiotic processes by which oil is imbued with value within petrocapitalism . . . the symbolic forms that rearrange daily experience around oil-bound ways of life."[31] It is a point echoed by Frederick Buell in his overview of the framing cultural and social narratives of the fossil-fuel era: for him, "[energy] (especially oil) remains an essential (and, to many, *the* essential) prop underneath humanity's material and symbolic cultures."[32] Might we thus read all of modern literature as a petroliterature, and so, too, make energy a necessary component of literary analysis?

The literary critic Graeme Macdonald has offered answers to these questions in his provocative and compelling overview of the key issues that have arisen for critics as they have begun to grapple with the en-twined histories of energy and culture. For Macdonald, "fiction, in its various modes, genres, and histories, offers a significant (and relatively untapped) repository for the energy-aware scholar to demonstrate how, through successive epochs, particularly embedded kinds of energy create a predominant (and oftentimes alternative) culture of being and imagining in the world; organizing and enabling a prevalent mode of living, thinking, moving, dwelling, and working."[33] His article unfolds the implications of what fossil fuels have meant for being and imagining the world, and what they have meant, too, for the fictional forms that have developed in relation to this world shot through with "narrative en-ergetics" and "psycho-social dynamics" linked to oil's force and power.

Macdonald argues that the inclusion of energy in literary analysis allows us to better grasp the ways in which distinct forms of energy help to make possible distinct modes of being and imagining. It is thus tempting—and the results are interesting enough to try it out—to map specific moments of literature in relation to developments in energy history. Modernism can be read in relation to the dissociation that is a central part of the history of fossil-fuel use, identifying in the key dates of 1870 (the moment when more energy is extracted from fossil fuels than from photosynthesis) and 1890 (the year in which more than half of global energy comes from fossil fuels) significant moments in the history of energy with resonances in literature and culture. Or one might consider postmodernism in relation to the 1973 OPEC crisis, a moment in which narratives of petromodern futures and US hegemony were deeply unnerved, and so, too, the self-certainties of narrative and the Western subject. Or to read postcolonialism, too, as an affirmation of a new era of colonial power linked to the fact that the moment of decolonization was also one in which the mantle of the biggest producer of fossil fuels was being passed from the United States to Saudi Arabia, Venezuela, and other producers in the Global South. To do so would undeniably enrich the vocabulary of literary and cultural analysis, and do so in a manner that forces us to recognize the significance of fossil fuels in the shape taken by modernity *and* to ask questions about how and why energy has for too long been a "nothing" in our assessments.[34]

But while we might thus enrich the critical literary vocabulary, the addition of energy here hasn't truly added an analytic element that helps us to better understand it. In each of the above examples, the work of analysis confers the socio-cultural significance of developments in the world of energy by indexing it to changes we already know about: re-configurations of energy did not produce modernism, nor create postmodern style, nor bring about the end of colonialism—which is not to say that they were insignificant. The challenge as we begin to figure energy into literary and cultural analysis is to know how to explain the precise social impact of a force whose significance is indexed in part by the fact that in much of the culture produced over an extended period, its import was not figured in culture. Macdonald is right: all of modern culture is a petroculture; we might also say that world literature is a world petroliterature. But in identifying world literature as the literature of an oil era, are we doing anything more than asking that we now pay attention to the ways in which literature can be used as a space that can be mined for insight into the character of this petro-era, presumably for other purposes, i.e., so that we might upend petroculture and bring it to an end?

Ian Baucom (2014) has written:

Although I have for some time accepted the force of Fredric Jameson's dictum that 'we cannot, not periodize,' until very recently it would not have occurred to me that postcolonial study, critical theory, or the humanities disciplines in general needed to periodize in relation not only to capital but to carbon, not only in modernities and post-modernities but in parts-per-million, not only in dates but in degrees Celsius.[35]

Is it productive or meaningful to undertake such a periodization? One could quibble with Baucom's choice of carbon or temperature as the measure of a new periodization, while still understanding the rationale for it: to re-imagine literature and culture in relation to environmental concerns.[36] A periodization developed around energy and forms of fuel use captures something that Baucom's proposed periods could never do: changes to social capacities and possibilities, to the ways in which access to energy—or equally, lack of access it—creates a predominant way of being and imagining that we usually describe as being modern. Despite all of the caveats, qualifications, and warnings I've provided about reading world literature as a petro-literature—as a literature that has to exist within a system organized around the capacities, fantasies, desires, and the imaginative and physical possibilities of oil—reading energy into literature and into the world *does* provide us with critical, political resources we might otherwise lack, even if care must be taken with precisely how we figure literature in relation to energy.

There is one final connection between energy and literature that needs to be addressed. I asked at the outset: does including energy in literary studies unnerve not just its *how*, but its *why*? Environmental historians J.R. McNeill and Peter Engelke describe the period since 1945 as the "Great Acceleration"—a period that "unfurled in the context of a fossil fuel energy regime and . . . exponential growth in energy use." It is also, they point out, "the most anomalous and unrepresentative period in the 200,000-year-long-history of relations between our species and the biosphere" and a period that is unlikely to continue.[37] Shouldn't we see literature as a key aspect of this period of Great Acceleration—as a practice that not only gives us insight into the period, but also a category and an activity (writing, read, and interpreting) that is an outgrowth of fossil-fuel use? I'm not suggesting here that literature is synonymous with the fossil-fuel era—a claim that comes across as absurd given the historical range of texts studied in any literature department. What I'm proposing is that we think more seriously about the links between the expansion of socially available energy and the coming-into-being of a general social capacity for literary activity—i.e., a segment of mass culture that emerged with expanded force and power at the turn of the twentieth century just as fossil fuels became more abundant (more energy for more people means more culture: a missing piece of the puzzle in Thorstein Veblen's assessment of the rise of mass culture). There is a largely unarticulated, widely-accepted, liberal narrative of literature, which links it to a developmental account of social capacity: more literature means more narrative which means more freedom and more social possibilities—as if world literature were, in the end, a policy of the World Bank.[38] One of the chief insights we gain by adding energy to world literature is that it allows us to see that all culture is petroculture because our understanding of culture is linked to oil's initial promise of unending abundance. It is always supposed to be better for there to be ever more culture, ever more energy. In recognition of modern culture's deep ties to energy, the link between literature and energy that critics and theorists have begun to make with ever greater frequency raises questions about the material and ecological weight of literature itself. To date, literature has avoided imagining its relationship to energy through units such as watts/hr or kilojoules/year, seeing itself as a medium that generates no environmental burden even as it proliferates awareness about the catastrophe of climate change. Figuring this meta-relationship of literature to energy might well generate an important and original ecological relationship to the apparatuses and objects of modern petroculture, and it might do so in a manner that would help undo the damaging self-certainties of the culture of liberal capitalism as well.

NOTES

1. I recognize that energy is not a physical thing, but a force—a property of objects that cannot be created or destroyed. As is common practice in English, I use energy as a substitute term for fuels, i.e., as an umbrella term to capture all of the fuels used today.

2. In "On Ideology," the final chapter of *On the Reproduction of Capitalism*, Louis Althusser writes:

    the subjects 'go,' they recognize that 'it's really true,' that 'this is the way it is,' not some other way, that they have to obey God, the priest, De Gaulle, the boss, the engineer, and love their neighbor, and so on. The subjects go, since they have recognized that 'all is well' (the way it is), and they say, for good measure: *So be it!* (197).

For Althusser, what makes subjects "go" are the Ideological State Apparatuses, which make them "go all by themselves" without (for the most part) the need of Repressive State Apparatuses.

In this same chapter, Althusser writes: "One can say that every social formation 'functions on ideology,' in the sense in which one says that a gasoline engine 'runs on gasoline'" (200). When I say that energy makes society and subjects "go," the connection to Althusser is deliberate. While I don't have the space to elaborate this here, it is worth drawing attention to the way in which fossil fuels are like ideology—absolutely essential to the operations of the social, but also largely invisible to their operations. A missing element in accounts of social reproduction to date has been an analysis of the process by which energy has come to be an element of ideology. "The subjects go, since they have recognized that 'all is well' (the way it is)" (197), even if what makes them go is quickly bringing the environmental systems in which they live to an end. See Althusser, *On the Reproduction of Capitalism*.

3. See, for instance, Barrett and Worden, eds., *Oil Culture*; Szeman and Boyer, eds., *Energy Humanities: An Anthology*; Szeman, Wenzel, and Yaeger, eds., *Fueling Culture: 101 Words for Energy and Environment*; and Wilson, Carlson, and Szeman, eds., *Petrocultures: Oil, Politics, Culture*.

4. J.R. McNeill and Peter Engelke point out that by 1870, human beings were already using more energy from fossil fuels than the annual amount of energy produced by all photosynthesis on the planet; since 1860, there have been *one trillion* barrels of oil used. McNeill and Engelke, *The Great Acceleration*, 8.

5. For an account of techno-utopianism, see chapter 4.

6. Smil, "World History and Energy," 559.

7. Smil notes that "only a few coastal societies collecting and hunting marine species had sufficiently high and secure energy returns (due to seasonal migrations of fish or whales) such that they were able to live in permanent settlements and devote surplus energy to elaborate rituals and impressive artistic creations (for example, the tall ornate wooden totems of the Indian tribes of the Pacific Northwest)." At a minimum, art as a social practice necessitates "surplus energy," which means that varying levels of surplus energy, in conjunction with broader changes in society animated by shifting forms of energy, would produce distinct modes of art practice as well as altering the social significance of art. In all of the changes that Smil narrates in relation to energy, his identification of art as "timeless" speaks more to his own unwillingness to figure art in relation to energy than to art's apparent ability to sidestep the historical shifts generated by changing forms and levels of energy. Smil, "World History and Energy," 550.

8. On cinema and energy, see Bozak, *The Cinematic Footprint*.

9. Malm, *Fossil Capital*. Though there is a tendency to imagine societies as having evolved "naturally" to ever more energy intensive systems and infrastructures, studies of the transition to fossil fuels highlight the political and social struggle that took place to make this happen. See Johnson, *Carbon Nation*; Jones, *Routes of Power*; and Malm, *Fossil Capital*.

The most acute analysis of the significance of energy transitions on politics remains Timothy Mitchell's *Carbon Democracy*. Mitchell describes the emergence of modern forms of mass democracy in conjunction with the use of coal: striking workers could easily impede the energy source on which nineteenth-century economies depended by blocking the rail lines that led from mines to factories and cities. In turn, the emergence of pipelines transporting oil has impeded such mass actions, while helping to animate the biopolitical practices of contemporary nation-states. See Mitchell, *Carbon Democracy*.

10. Yaeger, "Literature in the Ages of Wood, Tallow, Coal, Whale Oil, Gasoline, Atomic Power, and Other Energy Sources."

11. Jones, "Petromyopia."

12. Edward Renshaw points out "animals contributed 52.4 per cent of total work output in the United States in 1850; human workers, 12.6 per cent; wind, water, and fuel wood, 27.8 per cent; and fossil fuels, 6.8 per cent. In 1950, work animals are estimated to have contributed only 0.7 per cent of total work output; human workers, 0.9 per cent; wind, water, and fuel wood, 7.8 per cent; and fossil fuels, 90.8 per cent." Renshaw, "The Substitution of Inanimate Energy for Animal Power," 284. Thanks to Jeff Diamanti for bringing this article to my attention.

13. Christophe Bonneuil and Jean-Baptiste Fressoz point out that "If, in the twentieth century, the use of coal decreased in relation to oil, it remains that its consumption continually grew; and on a global level, there was never a year in which so much coal was burned as in 2014." Bonneuil and Fressoz, *The Shock of the Anthropocene*, 101.

14. According to the International Energy Association, in 2013 Canadians used 7,202 kg of oil equivalent per capita; by comparison, Germans used 3,868 kg of oil/ capita. The German figure represents a *decrease* from 1971 levels. See World Bank, "Energy Use Per Capita," *World Development Indicators*, http://data.worldbank .org/indicator/EG.USE.PCAP.KG.OE (accessed August 17, 2018).

15. For a recent account of the emergence of a new fuel in Madagascar—charcoal—see Norimitsu Onishi, "Africa's Charcoal Economy is Cooking. The Trees are Paying," *New York Times*, June 25, 2016, http://www.nytimes.com/2016/06/26/world /africa/africas-charcoal-economy-is-cooking-the-trees-are-paying.html?_r=0 (accessed August 17, 2018). According to the United Nations Food and Agricultural Organization, over the past twenty years charcoal production in Africa has doubled; today it accounts for 60 percent of total global production. The UN also predicts that demand for charcoal in Africa will double or triple by 2050, in large part as a result of population increases on the continent.

16. Ghosh, *The Great Derangement*.

17. The dates are about right: Goethe claims in 1827 that a world literature is "beginning," while Marx speaks famously in 1848 in the *Communist Manifesto* of a "world literature" arising out of the "the many national and local literatures." If these comments predate the oil era, they speak to the capacities and possibilities that are emerging as coal begins to be used to generate an increasingly larger part of the energy used in Europe and the UK. In their "Introduction" to *Materialism and the Critique of Energy*, Brent Bellamy and Jeff Diamanti argue "Marxism's methodological and theoretical development in the 1840s occurs not despite but within

the contemporaneous surfacing of the theory of energy across Britain, Prussia, and France ... Marxism's historical materialism discovers itself in the industrialization of energy, and energy therefore was and remains central to historicizing the fissures of a planet reverberating against the fault lines of an exhausted stack of capital today." Bellamy and Diamanti, *Materialism and the Critique of Energy*, ix.

18. Klare, "Hooked! The Unyielding Grip of Fossil Fuels on Global Life."

19. Brown and Szeman, "What is the Multitude?", 109.

20. Mitchell, *Carbon Democracy*, 109–43.

21. See, for instance, Colebrook, "Not Symbiosis, Not Now: Why Anthropogenic Climate Change Is Not Really Human" and Haraway, "Anthropocene, Capitalocene, Chthulucene: Making Kin." Critics of the concept of the Anthropocene are far more common than defenders of it; one exception is Dale Jamieson's argument for an "ethics" that emerges from the concept of the Anthropocene. See Jamieson, *Reason in Dark Time*, 185–92.

22. Moretti, "Conjectures on World Literature," 56.

23. Moretti, "Conjectures on World Literature," 65.

24. Quoted in Moretti, "Conjectures on World Literature," 56.

25. Wenzel, "Petro-Magic Realism," 456.

26. See Ghosh, *The Great Derangement*.

27. In *Cyclonopedia*, oil possesses qualities well beyond those we normally assign to it. Negarastani describes it as nothing less than a "satanic sentience" that "possesses tendencies for mass intoxication on pandemic scales (different from but corresponding to capitalism's voodoo economy and other types of global possession systems)." Negarastani, *Cyclonopedia*, 26.

28. Jones, *Routes of Power*, 143.

29. Munif, *Cities of Salt*, 222.

30. The most compelling account of the impact of energy on literary form remains "The Work of Art and the Work of Nature at the Dawn of Oil," the first chapter of Jeff Diamanti's PhD thesis, "Aesthetic Economies of Growth: Energy, Value, and the Work of Culture After Oil." See Diamanti, 2015, 72–117.

31. Barrett and Worden, "Introduction," in *Oil Culture*, xxvi.

32. Buell, "A Short History of Oil Cultures," 70.

33. Macdonald, "The Resources of Fiction," 4.

34. Macdonald writes: "if *all* fiction is potentially energetic, valorizing energy use, then how do we kinetically assert our claims and configure our readings to make it more apparent?" Macdonald, "The Resources of Fiction," 19.

35. Ian Baucom, "History 4°: Postcolonial Method and Anthropocene Time," *Cambridge Journal of Postcolonial Literary Inquiry* 1, no. 1 (March 2014): 125.

36. "A carbon focus is reductionist, possibly the greatest and most dangerous reductionism of all time: a 150-year history of complex geological, political, economic, and military security issues all reduced to one element—carbon." Princen, Manno, and Martin, "The Problem," in Princen, Manno, and Martin, eds., *Ending the Fossil Fuel Era*, 6.

37. McNeill and Engelke, *The Great Acceleration*, 2, 5.

38. See Kumar, ed., *World Bank Literature*.

CHAPTER 11

_____

Pipelines and Territories: On Energy and Environmental Futures in Canada (2018)

I want to provide an analysis of the politics of pipeline expansion and development in Canada since the beginning of the twenty-first century. My aim in doing so is not only to outline the vexed drama of pipelines in Canadian political life, but also to see what a focus on pipelines might teach us about the politics of oil and the environment at the present time, and what it might tell us, too, about the energy futures into which we are moving.

The existing system of oil and gas pipelines in North America is massive. There are more than 840,000 km (522,000 miles) of pipelines in Canada alone, which includes 117,000 km (72,700 miles) of large-diameter transmission lines of the kind on which I'll be focusing.[1] In addition to the pipelines that crisscross each country on the continent, there is also a significant network of cross-border pipelines, including major pipeline networks built as recently as 2010 that cut across the US/Canada border. While pipelines have drawn media and political attention in the past—most notably, in Canada, in relation to the Mackenzie Valley Pipeline Proposal and the Berger Inquiry—the past several years have seen pipeline construction become front-page news in Canada and the United States.[2] Why? What has changed about pipelines that has made these deliberately invisible forms of infrastructure now visible? And in turn, what does this new visibility say about environmental and cultural politics in relation to infrastructure such as pipelines?[3]

I approach these questions by first surveying the politics of three major pipeline projects that are pressing issues of public policy in

Canada—the Northern Gateway, Energy East, and Trans Mountain Expansion pipeline projects—and a fourth that has made headlines in Canada, the United States, and around the world: the Keystone XL pipeline project, which was killed in November 2015 by an executive decision of President Barack Obama and revived by President Donald Trump in 2017.[4] I follow this with an exploration of *pipeline theory*—an elaboration of the physical and conceptual dynamics of this particular form of infrastructure and the politics that it calls into being—before concluding with an all-too-brief (but hopefully suggestive) discussion about the role that pipelines, in their sheer materiality and new visibility, play in politics today—and not just politics of the environmental kind.

A. Border Crossings: The Politics of Pipelines

In January 2016, the new government of Canadian Prime Minister Justin Trudeau announced that the federal government had decided to mandate additional reviews to assess the environmental impact of new pipeline projects. The government was also instituting a requirement that pipeline companies engage in further consultations with First Nations communities about all of the major pipelines currently being proposed, and offered funding to these communities so that they might expand community consultation on the Energy East pipeline project in particular.[5] This announcement of new environmental reviews came on the heels of a report issued by the office of the federal environment commissioner, which severely criticized Canada's National Energy Board (NEB) for its failure to track whether companies actually meet the conditions set out in the approval process.[6] In the past, the NEB had been repeatedly criticized for giving an easy pass to energy projects; now it would have to devote more time and energy to making sure that new projects meet criteria that an environmentally conscious public was demanding of infrastructural projects.

For most in a country that depends on natural resources (and fossil fuels in particular), these announcements by the Trudeau government came as a surprise. For several years, the large decline in the price of oil had meant far fewer dollars flowing into both federal and provincial government coffers (as of May 2018, the price of a barrel of West Texas Intermediate crude was about $75 CAD—higher than it had been for years, but still below its price in January 2015, when it was over $100 per barrel). In February 2016, the federal Liberal government announced a projected budget deficit of $30 billion dollars for the coming year—a huge deficit in Canadian terms and triple the already large deficit that the Liberals had anticipated for 2016 at the end of 2015. The Liberals also announced significant transfers of federal funds to the province of Alberta, the largest producer of oil and gas in the country, and the province most deeply impacted by the decline in oil prices. The Trudeau government's announcement of additional environmental assessments extended substantially the length of time it takes for pipeline assessments—or at least in theory. In the case of Energy East, for instance, it was estimated that it would increase the assessment time by at least half—if, that is, the NEB is properly able to undertake its new mandated task at all: measuring the impact of the new pipelines on greenhouse gas emissions in Canada, both in the present *and* in the future.[7] Those clattering alarm bells the loudest were, of course, industry leaders who, if not opposed to the new assessments altogether, were questioning the timing of their imposition given present economic circumstances for the industry. Those applauding the Trudeau government's decision included not only the environmental community and First Nations, but also political leaders in British Columbia and Quebec, including the mayors of many of the communities through which proposed pipelines would cut; these were the same groups and individuals who had been most critical of the federal government as it reversed its original hardline stance, specifically in the case of Trans Mountain Expansion, which was approved in January 2017.

If pipelines have become one of the signal issues in contemporary Canadian politics—now less a surprising political hot topic than one treated by media and public alike as a matter of obvious import—it is because these infrastructural assemblages capture the key political anxieties, divisions, and struggles shaping the country at the present time. These tensions and struggles can be grasped in two editorials written days apart in the *Globe and Mail* early in the Trudeau government's first year. On January 24, 2016, the *Globe* applauded Prime Minister

Trudeau's expansion of environment assessments on pipeline projects, while also insisting

> It would be a huge mistake on his part to fail to sell the merits of Energy East. . . . Mr. Trudeau needs to persuade Canadians of the fact that a healthy energy sector is a key part of a healthy economy, and of the consequence flowing from that: Oil must move. Where pipelines can transport oil safely, efficiently and in an environmentally respectful way that passes muster with a timely, arm's-length review process, they should be built.[8]

A week later, the same newspaper urged the new PM to take a leadership role in getting Canadians on track with the policy decisions the country will need to take if it is going to address levels of greenhouse gases: "to produce the kind of sharp drop needed between now and 2030, Canada will have to amputate, not nip and tuck."[9] This is, to say the least, a mixed message: Canada needs to get serious about its economy, and so it needs pipelines; Canada needs to get serious about its environment, and so more pipelines moving more oil might be the last thing it needs. In the drama of pipeline politics, Canadians are encountering the political schizophrenia produced when capitalist societies try to address global warming through the same mechanisms that generated the condition in the first place. The capital generated by the oil sands is needed, it is often argued, in order to create new technologies through which the consequences of oil sands production and consumption can be addressed. The obvious contradictions of such a view has meant that it has been easier for publics to deal with one part of the issue at a time—as with the *Globe* editorial, fully in support of pipeline expansion on Monday, fully in support of strong environmental policies on Friday. Nothing in the period since these editorials were written has altered this slip-and-slide, back-and-forth commitment to both the environment *and* pipeline and energy projects—an indictment of the business-as-usual approach to climate change that has characterized the programs of governments around the world. Indeed, the failure of the Trudeau government to live up to its initial obligations to an expanded consultation process, and its backsliding on commitments to communities and First Nations, points to the power that resource money continues to have on politics in Canada.

Pipelines are a matter of geographic and economic necessity everywhere they are found, but especially in Canada. Alberta is the major site of oil production in Canada; the oil sands have been estimated to contain 178 billion barrels of economically recoverable oil (the third largest known oil reserve on the planet after Saudi Arabia and Venezuela). Alberta is also one of only two landlocked provinces in Canada. Without a port of its own through which to supply oil to global markets, the province has only ever been able to supply a single customer—the United States (Canada is the largest single supplier of foreign oil to the United States; Canadian production in 2014 was 3.535 million barrels per day, of which 3.388 million went to the United States).[10] The pipeline projects that have generated public debate and discussion in Canada are owned by different companies and travel different routes. However, all have the same ambition: to move oil from Northern Alberta to coastal ports so that it might find new markets and higher prices. The four-phased Keystone project (three phases of which are complete, including a cross-border pipeline running from near Edmonton, Alberta, to a tank farm in Patoka, Illinois) is intended to drag Canadian oil all the way to the refineries of Port Arthur, Texas. Enbridge's Northern Gateway project pipeline (which is, as of 2018, apparently dead) was designed to connect the oil sands up with the growing Asian market for energy, and would run from just north of Edmonton to the north coast of British Columbia.

Other pipeline projects have also made news in recent years: Kinder Morgan's Trans Mountain Expansion project and TransCanada's Energy East. Trans Mountain is a proposed expansion of an existing pipeline that runs from Edmonton to Burnaby, British Columbia. As an expansion of an existing pipeline entails a quicker and less stringent review process, Trans Mountain anticipated relatively little difficulty in getting the project approved, becoming in the process the company to have hit the oil sands export jackpot. Unfortunately (for Kinder Morgan, at least), the pipeline ends in one of the most environmentally sensitive regions of the country: the Lower Mainland of British Columbia. Kinder Morgan's proposed project has produced a ferocious response, including protests on Burnaby Mountain where Simon Fraser University is located; these led to threats of imprisonment for some of the lead protestors, including poet and professor Stephen Collis.[11] The city of Burnaby has also challenged the authority of the National Energy Board to make decisions about the pipeline project on its behalf, and the city's mayor, Derek Corrigan, has stated that he would be willing to end his career by getting arrested while attempting to stop the pipeline.[12] Despite this and other resistance to the project, it was approved by the Trudeau government in January 2017. Even in the wake of the approval, resistance to the expansion remains active. In 2018, the decision by the newly elected government of British Columbia to cancel the Trans Mountain pipeline has been actively opposed by the federal government and the Alberta government, and has made the front pages of newspapers across Canada.

The pipelines under discussion are of incredible length: the proposed Keystone XL pipeline is almost 1900 km (1181 miles) long, while the completed first phase of the Keystone project is nearly 3500 km (2175 miles) in length. Even this massive pipeline is dwarfed by a second proposed TransCanada project: Energy East, which would extend from Alberta to refineries in Montreal, Quebec City, and Saint John, New Brunswick (at this last site, Irving Oil has announced plans for a $300-million refinery terminal that would employ oil delivered by Energy East). When completed, the $15.7 billion Energy East would be the longest pipeline in North America and one of the longest in the world.

Even after the Trudeau government's recent warming toward the oil industry, Energy East still faces delays due to the new environmental regulations being imposed by the federal government. It also faces many of the same blocks and limits that the other pipeline projects have faced. Many affected communities, including Kenora and North Bay, Ontario, oppose it, and former Montreal mayor, Denis Coderre, publicly challenged the merits of the project. Environmental groups have also been aggressive and proactive in their opposition to Energy East. To give just one example, in mid-February 2016, a coalition of environmental groups filed a motion against TransCanada pipelines in Quebec Superior Court, claiming that the company had failed to file a project notice with the province's Environment Department; the filing of such a notice would trigger an environmental impact assessment at the provincial level. Finally, as with the other pipeline projects I've mentioned, Energy East faces challenges by many First Nations communities. By remaining in Canada and staying away from the pristine West Coast, it appears that supporters of Energy East in government and industry had hoped to sidestep some of the issues that have plagued other pipeline projects. But the length of the Energy East pipeline comes with its own problems, including the fact that it will cross 180 First Nations territories. There are a huge number of groups and communities to appease before oil starts traversing the length of the country—an industrial behemoth as symbolically expressive of the ties that bind Canada in the twenty-first century as the railroad in the nineteenth and the TransCanada Highway in the twentieth.

B. Pipeline Theory

Despite the various actors involved in each of the projects I've described, the fundamental issue at work in all of them is the same. What shape should our energy future take? Are Canadians going to continue using fossil fuels to fuel their economies in the short and (perhaps) long term? Or can we imagine a shift in our use of fossil fuels that might lead to a real change in the impact of hydrocarbons on the environment—not a partial shift, a temporary shift, or a haphazard shift, but a move that says: we're done with the fossil-fuel era and all the imaginaries it has birthed? Whether it has been governments, First Nations communities, academic communities, or environmentalists that have challenged Canadian pipeline expansion, the focus of their efforts has been to attract public attention to its environmental and social impact. For those who have opposed its expansion and extension, more pipeline infrastructure in Canada can't help but lead to greater CO_2 emissions. Even without these massive projects, Canada is poised to generate 857 megatons of CO_2 by 2020, missing the modest mark of 611 megatons of CO_2 by 2020 that the country set for itself in advance of the 2015 COP21 meeting in Paris. Without a significant change in how the country imagines its relationship to fossil fuels, Canada will become the worst of the OECD countries as measured by its ability to reach its climate targets.[13] In this respect, at least, the new Trudeau government looks to be little different from the former government of Stephen Harper. Despite the fact that it is more conscious of and alert to the reality of climate change and the necessity of developing policy to address it, it has done little to move Canada away from fossil fuels, other than to propose a carbon tax that is both regressive and ineffectual. Indeed, in the wake of a decade of aggressive Conservative politics, the new federal Liberal government and the presence of NDP governments in Alberta and British Columbia have tended to ease the fears of the Canadian public about the environment—a disaster in the making, which future governments will be hard pressed to set aright.

As I have already indicated, pipelines have become a symbolic site to enact the broader politics of the environment in Canada; they have come to play a significant role in the United States, too, as a result of struggles over Keystone XL and protests at Standing Rock over the Dakota Access Pipeline. *Why* pipeline infrastructure has taken on this role is less obvious than it might seem. Throughout their history, oil pipelines have (with some notable exceptions) aspired to be invisible. As Darin Barney reminds us, "just as it is best when digital networks deliver us images, sound, and text wherever and whenever we want them without bothering us, it is best (at least from the perspective of energy capital, energy states, and energy consumers) when pipelines deliver energy without anybody noticing them."[14] Dull, dead, apparently meaningless rods of steel and plastic, pipelines are *everywhere*—a vast capillary network linking extraction sites to fuel terminals, terminals to refineries, and refineries to factories, businesses, and homes. The most significant feature of this massive system is that it has managed to remain invisible even as it has expanded to supply increasing levels of product to an ever-expanding economy and population. The majority of Canadians would have no idea where to find a supply pipeline in their community, and would find it even harder to point to the massive cross-border pipes that hold the system together; it's no coincidence that those for whom pipelines are all too visible—those in impoverished or rural communities, or indigenous Canadians whose territories pipelines often crisscross—have until recently had their voices ignored in relation to decisions by government and industry.

This invisibility does not mean that pipelines are insignificant. Far from it. A map of the major pipelines moving across and around a country highlights concentrations of power, money, and influence linked to resource extraction. Such a map would show, too, spaces and sites of danger to human communities and ecosystems. According to Natural Resources Canada, "between 2009 and 2013, 99.999 percent of the crude oil and petroleum products transported by Canada's federally regulated pipelines arrived safely, and during the past three years (2011–2013), 100 percent of the liquids released by these pipelines were completely recovered."[15] Impressive enough numbers, yet the latter figure suggests something more ominous about the former figure than we are intended to garner. Given the size of Canada's pipeline system, a fail rate of 0.001 percent constitutes 825 km (513 miles) of pipeline problems—about the highway distance between Toronto and Québec City (or Toronto and Chicago), and sure to include waterways and threatened ecosystems. Their increasing visibility when they fail—and as the pipeline system ages, it has come to fail more and more—has drawn attention to the scale of their presence. The possibility that any aspect of the pipeline system might fail at any given time has transformed the entire system into a looming threat. In recent years, the pipeline system's importance or necessity for modern societies has been questioned as it has emerged into sight—an inversion of the desired function of its hiddenness that is key to understanding the reanimated politics surrounding pipelines.

Barney has noted that pipelines have to be read as "media in, with and through which we come to be in the world as the sort of beings we are."[16] It is through such media that social reality is designed, built, and organized. As surely as other forms of media, pipelines generate meaning through their very material existence, but also act as conduits of symbolic meaning making:

> State approval and regulatory processes for pipeline developments are media for the production and circulation of contested scientific, technical, economic, and political knowledges about what pipelines are, what they do, and what they mean. Studies are made, presented, contested, and archived. There are blooms of data and information. Discourses are mobilized, claims are made, and languages

are translated. State and corporate public relations machines are swung into high gear. There are demonstrations, occupations, and protests. Moving and still images, graphics, text, voice, and sound proliferate via a similarly diverse array of media that together comprise a network of which the pipeline-to-come forms the trunk. Almost none of this activity would be possible without petroleum and the pipelines that communicate between its source and many destinations.[17]

The characteristic invisibility of pipelines has long served an ideological function. The surfeit of symbols that Barney names points directly to the use of resources for the benefit of capital and the garnering of massive profits through the application of technology to property. For industry, demonstrations, occupations, and protests are thus to be kept in abeyance; to be maximized are the profits that flow into bank accounts as surely as oil flows through pipelines, and to this end these media function best if made invisible.[18]

Pipelines have from their origin been systems of power and money. In his nuanced and compelling account of the history of the development of pipelines in the United States, Christopher Jones notes that "pipelines were not simply a mechanism for moving oil; they were an explicit attempt to transform who controlled the flows of petroleum and who would profit from them."[19] One of the primary reasons for the dominance of John D. Rockefeller's Standard Oil Company in the early decades of the US oil industry was the preferential rates he had established with railroads for shipping oil from the hinterlands, where it was extracted, to refineries on the US east coast. Pipelines emerged as a device through which a new oil company—Tide Water—was able to circumvent Rockefeller's control of the railroads, and so move its oil at more competitive (even cheaper) rates than could be managed by Standard. From the very beginning, the need for pipelines to traverse large territories was one of their major limitations. Rockefeller tried to impede Tide Water's attempts at creating pipelines by buying up property that lay in their path and by working to block them from crossing railroad lines. Even given the many obstacles generated by Standard, Tide Water inaugurated the world's first long-distance oil pipeline in 1879; by 1884, more than three-quarters of crude oil distribution had shifted from rail to pipelines.[20] The consequence of the mass introduction of pipelines was significant. By reducing costs, the creation of pipelines intensified the consumption of oil. In its early years, oil was used primarily for lighting. However, the low cost of shipping it via pipeline created new markets for the use of oil for heat and power. The rapid adoption of oil as the primary energy source for an expanding modernity—an expansion it helped fuel—solidified oil as the key resource for the operations of capitalism, and a resource that generated an enormous return on profits even when the price of oil was (relatively speaking) low.

Pipelines have acted as mechanisms of power in at least two other ways. First, according to Timothy Mitchell, "oil pipelines were invented as a means of reducing the ability of humans to interrupt the flow of energy."[21] In Mitchell's account, the shift from coal to oil interrupted and dislocated a form of political protest that had developed alongside the rise of coal production. The ability of coal miners to effectively and immediately disrupt energy flow through mass strikes or sabotage gave their political demands special force and led to major gains for workers between the 1880s and the interwar decades, while also supporting the development of workers' consciousness of their social circumstances. All that was required for political action was for workers to block the railroads through which coal made its way from extraction sites to communities and cities that had grown dependent on the fuel. "Unlike the movement of coal," Mitchell reminds us, "the flow of oil could not be readily assembled into a machine that enabled large numbers of people to exercise novel forms of political power."[22] Second, pipelines are a dissociative mechanism that frames labor in relationship to fuel in a distinct way. Compared to the extraction of coal, oil requires far fewer laborers per unit of energy, and pipelines ensure that the fuel extracted is used at a distance from the origin site.[23] This dissociation of extraction from consumption has implications for the environment as well as for labor. Jones points out that from the very beginning of the establishment of the pipeline system, "the users of oil gained the benefit of cheap energy without assuming responsibility for its environmental damage."[24]

Over the twentieth century, pipelines generated further socio-political and environmental dissociations as they shifted from identifiable mechanisms of power and control (over which there might be struggle) to a rationalized, techno-scientific process of resource production (over which there isn't struggle). In her account of the construction of the Trans-Arabian Pipeline, Rania Ghosn emphasizes the way in which the erasure of geography by pipelines "abstracts technological systems—their materialities, dimensions and territorialities. It removes from representation the territorial transformations along the conduit, which the inscription of the infrastructure produces, and overlooks the politics of consensus or dissensus necessary to distribute resources."[25] In this, the pipeline followed the path taken by scientific processes in general over the course of modernity: technology as *Gestell,* an enframing of the world in which nature becomes a "standing reserve" that underwrites the deadened quest of capital to fuel its own drama (Heidegger).[26] As a technological process, a standard tool used by industry scientists, and a practice taught uniformly at engineering schools across the world, pipelines have become a prime example of what Keller Easterling has named "extrastatecraft." "Contemporary infrastructure space is the secret weapon of the most powerful people in the world precisely because it orchestrates activities that can remain unstated but are nevertheless consequential," Easterling writes. "Some of the most radical changes to the globalizing world are being written, not in the language of law and diplomacy, but in these spatial, infrastructural technologies—often because market promotions or prevailing political ideologies lubricate their movement through the world."[27] For Easterling, contemporary infrastructure—and by this she means everything from free trade zones to broadband media protocols to ISO global management standards—has an agency, capacity, or disposition through which it exerts power, both separate from and in partnership with the actions of states. She is especially interested in the ways that infrastructure has been shaped to enable and support neoliberal capitalism. In short, extrastatecraft is neoliberalism carried out or enacted by the infrastructures of modernity. The power of extrastatecraft lies precisely in the fact that infrastructures tend to be seen as neutral, rational, and technical solutions to modern problems, and so are seen, too, as devoid of political interest or impact. To the physical invisibility of pipelines, we can add what we might describe as their "political invisibility." That is, pipelines function as actants that enable capitalism both through their operation and their technological rationality, which has redoubled their givenness as a system that we (supposedly) need and (supposedly) can't live without.[28]

What is remarkable about the current public character of the discussion over pipelines in Canada is not that they have become more physically visible (even at sites of protest such as Standing Rock, the actual physical apparatus of the pipeline remains hidden), but more *politically* so. As a result of the debates and disputes over their necessity and rationality, the worldview contained within the technology of pipelines has been exposed and the ideologies contained within it made open to challenge. Some of this new political visibility can be explained as a consequence of the development of a more intensified environmental consciousness on the part of publics and the greater inclusion of the environment in policy-making within the operations of official state politics. The new politics of pipelines in Canada and the United States would in this sense present an example of what Jacques Rancière has described as "the distribution of the sensible"—a shift in the "very configuration of the visible and the relation of the visible to what can be said about it."[29] One could point to a range of reasons for this reconfiguration, including oil spills across the world that have gnawed at the dissociative function of pipelines, images that have now circulated for decades drawing attention to the scale of the oil sands, and, in the Canadian case, the devastating explosion in 2013 of train cars carrying oil in the center of Lac-Mégantic, Quebec, which killed 47 people and destroyed more than 30 buildings. The actual physical apparatus of pipelines might be as invisible as always, pushed off to the hinterlands and to zones of private property; and yet, the logics of their operations and the world they bring into existence are now newly available to dispute and debate.

The emergence of a new pipelines politics is linked to the operations of a *dispositif* to which we have grown so accustomed that we have forgotten their power—that of borders. Pipelines are technologies that enact forms of extrastatecraft, gliding below the surface of state politics even while helping to sustain them. Easterling's presumption is that the technologies of extrastatecraft will always work in conjunction with statecraft, amplifying and accelerating the neoliberal logics of the latter. And for the most part, this may well be the case. The current struggle over pipelines in Canada, however, offers an example of when statecraft and extrastatecraft collide, with the result that the hidden demands and suppositions of infrastructure are revealed for what they are. Despite all of the opposition that has been mounted against it, the Keystone pipeline project linking Edmonton and Oklahoma is largely complete: three of the four segments are already done. The segment that was blocked—the XL segment—was one that crossed national borders and (at least for a time) brought infrastructure into collision with the imperatives of state. The rejection of the project by the Obama administration is an index of a shift in attitudes toward the apparent rationality of the fossil-fuel era; the turnaround approval by the Trump administration constitutes little more than a furtive last stand of an oil-powered hegemon in a world that cares less and less for the fuel and the Cold War imaginaries it powers. This isn't to suggest that governments are ready to give up on fossil-fuel extraction entirely or that an environmental ethos now pervades halls of power and governs policy-making in relation to energy and natural systems. And yet, the very public struggle over phase 4 of the Keystone project only years after the untroubled approval of phase 1, and over projects such as the Dakota Access Pipeline and other Canadian pipelines, suggests a political shift to which it is necessary to attend.[30]

The borders involved in the Canadian pipeline projects I've been discussing extend beyond national ones. In the movement of oil to port, multiple sovereignties come into play even within Canada. To begin with, the Canadian federal government is at the center of these policy discussions only because the proposed pipeline crosses provincial borders. The Canadian constitution assigns control over natural resources to provinces, *not* the federal government; of the 840,000 km (522,000 miles) of major pipelines in Canada that I noted at the beginning of this chapter, only 73,000 km (45,360 miles) are under the mandate of Natural Resources Canada and the National Energy Board. To the federal government must be added the imperatives of individual provinces, cities, and finally, and perhaps most significantly, First Nations, who have been increasingly vocal and active about control and decision-making with respect to their territories. There are all manner of existing pipelines crossing provincial, federal, and territorial space, many of them built through the use of government expropriation of private land. Standard modes of political expropriation have been rendered ineffective in relation to the building of these pipelines, however, not only because of political difficulties in expropriating First Nations territory, but also because the varied sovereignties involved have distinct views on the environment and the function of the pipelines in relationship to environmental futures.[31]

The transition of pipelines from mechanisms of extrastatecraft to objects of statecraft has prompted a struggle over national futures in Canada—especially national-environmental futures, the two terms now indelibly linked in relation to political decision-making on any and every topic. Despite the fact that all of these pipelines are projects of individual companies that benefit their bottom lines, the importance of these pipelines for the purposes of national unity has been echoed by government leaders as much as by business executives. "Will the prime minister pick up the phone, call his friend, the mayor of Montreal and tell him to smarten up and start standing up for Canadians all across Canada?" Conservative critic Candice Bergen demanded of Trudeau while the prime minister was in Davos in 2016.[32] The Conservative opposition leader, Rona Ambrose, has claimed those opposed to large pipeline projects are generating a crisis of unity[33]; the rhetoric of "national unity," "nation-building," and nationalism is being circulated in ways that are new to a neoliberal Canada.[34] In the language used by the right, one of the major problems about the pipeline debate is that it politicizes what should properly be a technological, regulatory issue: the import of the pipelines is a "no brainer" (in Ambrose's words) that should be consigned to the invisible space of extrastatecraft. But even the need to make the case that pipelines should be invisible renders them newly political and available for investigation and interrogation by all those who might be impacted by them. The Harper government developed its legitimacy around the redefinition of Canada as an energy superpower. The ongoing pipeline debate in Canada might well end with federal approval over Energy East in addition to Trans Mountain and Keystone XL, with the rationale being to keep jobs in Canada and to improve the economy. However, in the process, there will have been a very public interrogation of the terms of national unity, the function of extrastatecraft like pipelines in shaping the field of debate, and finally, of the character of the country's energy futures and environmental commitments.

C. From Extrastatecraft to Statecraft: Toward an Energy Commons?

In his analysis of the complex politics of another pipeline project—the Baku-Tblisi-Ceyhan pipeline from Azerbaijan to the Turkish coast— Andrew Barry writes:

> Theorists of radical democracy have focused on the articulation of disputes between human collectives, the identities of which are shifting and relational. But . . . they have had less to say about the importance of materials and technologies in political life and how the properties and behaviour of organic and inorganic materials—whether they are diseases, climate change, animal species, mineral resources or new technologies—themselves participate in such controversies . . . [M]aterial objects should not be thought of as the stable ground on which the instabilities generated by disputes between human actors are played out; rather, they should be understood as forming an integral element of evolving controversies.[35]

In an earlier paper called "On Energopolitics," I argued that one of the limits of states with respect to global warming is connected to the nature of state power itself.[36] It's not only that state power is delimited geographically, while global warming takes in the space of the entire planet. Rather, state power has no concept of—and no relation to—either energy or the environment. While some have argued for a deep connection of Michel Foucault's articulation of state power with the environment, his theories of the constitution of subjects and of states, and of all the systems and mechanisms involved in producing and managing both, do not include *any* interest in natural systems and their limits.[37] Remember: Foucault's ideas on the organization of subjects and power have to be seen as analytic rather than normative accounts of power; we might want our state systems to be different than they are (i.e., to include the environment), but they haven't developed in this way, and so adding energy and the environment to the operations of power involves more than just hoping that states might attend to global warming. In her assessment of the challenge that climate change poses to our understanding of the operations of biopolitics, Hannah Knox argues that the concept of "population" that climate scientists are working with is "not a population constituted through a political project of statistical aggregation, but a rather 'empty' conceptualization of population that appears as the only available interpretation of the causes of a particular material effect."[38] The effect of this "empty population" on political action with respect to the environment is significant. If population has constituted the major site at which states configure power/knowledge and is also the principal guarantor of political authority, the "empty population" of environmental crisis constitutes "a new space of *not-knowing* with implications for the framing of practices of climate change governance."[39] Knox's revelation of the empty population used in environmental analysis reinforced my own conclusions in "On Energopower" about state power in relation to the environment: the fact that states are mechanisms for the organization of power and are relatively insensate to the environment and its populations means that we look in vain to them to address global warming, especially with the radical speed and at the radical scale necessitated by the problems at hand. States work only on defined populations; we have no extant political structure that speaks to the population of the planet as a whole, much less to the non-human species and objects with which we share the planet, and with whom we would need to shape a new planetary politics.[40]

The new political visibility of pipelines in Canada won't alter the constituent components of biopower. What it does do, however, is give us a better understanding of the gaps and limits of statecraft in relation to the environment, as well as a more thorough sense of the political pressures exerted by forces of extrastatecraft. Importantly, it may also have a function in reshaping the terms of the debate about what oil is for and *whom* it is for. In "The Petroleum Commons," George Caffentzis notes that while water has long been seen as a common property of communities, petroleum has always been owned, whether by magnates like Rockefeller or leaders such as Saddam Hussein. Might it be possible to imagine oil and other sources of energy as common property in the same way that we imagine water and air to be common? As something owned by Canadians *qua* Canadians (and indeed, humans *qua* humans), and not delimited by extractive rights or the vagaries of provincial boundaries established in advance of the fossil-fuel era, and certainly in advance of knowledge of its environmental consequences, which recognizes no borders? A decade ago, Caffentzis argued (too hopefully, it turns out) that a petroleum common was in fact slowly emerging, through indigenous claims to oil, the politics of social movements, and international organizations such as the United Nations. The proponents of a petroleum common, he writes, "argue that the consequences of the exploration, extraction, distribution, and consumption of petroleum are so problematic for 'humanity' that they cannot be left to the devices of private companies or nation states, but have to be managed by international organizations."[41]

The struggle over long-distance pipelines and the borders they cross has transformed a hidden aspect of the infrastructure of modernity into a space for the articulation of new demands and new desires for our energy futures; whether the product of the oil sands ever makes it to foreign markets now depends more on political struggles than on the technical prowess of forcing oil across a vast country now suddenly alert to the world that it has brought into existence. The debates over the merits of these pipelines index, perhaps, the beginning of a new political dynamic in relation to environmental futures, one in which the difficult changes that need to be made about infrastructure and power are visible as never before. This dynamic is importantly different than one might expect. A petroleum commons, such as the one imagined by Caffentzis, configures a state form appropriate to the empty population invoked in discussions of climate change. But Dipesh Chakrabarty has repeatedly cautioned us to be aware of the fact that there is a misfit between the politics of the state and the politics of the environment, between the "globe" of globalization and the "globe" in global warming.[42] In Caffentzis' petroleum commons, the two globes are flattened into one, resolving the problem of petroleum and of climate change by imagining a bigger state to manage a bigger population without difference or distinction.

The energy future promised by the new visibility of pipelines is different from this. In a space of struggle that brings together borders and nations, the cultural and material, and the claims and demands of distinct communities, the whole apparatus of modern politics and its environmental consequences is on display and open to challenge—not to be quickly closed off, but so that we might crack open the claims that modernity has made on us and shape a commons no longer made in its image, shaped by its expectations and beholden to the fuels that have for too long powered it. As the infrastructure of oil modernity becomes ever more visible, so, too, will the violence and exclusions of the oil capitalism to which this infrastructure has given shape. And so, too, will new political forms become ever more present and possible as we move deeper into this century.

NOTES

Thanks to Adam Carlson and Jordan Kinder, who engaged in research that helped bring this chapter together.

1. Natural Resources Canada, *Pipelines Across Canada*, May 2016, para. 1, https://www.nrcan.gc.ca/energy/infrastructure/18856 (accessed August 17, 2018).

2. See Berger, *Northern Frontier, Northern Homeland* and O'Malley, *Past and Future Land*.

3. On the visibility of oil infrastructures during the 1970s energy crisis, see Wellum, "The Ambivalent Aesthetics of Oil."

4. Adrian Morrow and Shawn McCarthy, "Trump moves forward on Keystone XL," *Globe and Mail*, Jan. 25, 2017, A1, A9.

5. Shawn McCarthy, "Ottawa adds additional steps to pipeline reviews," *Globe and Mail*, Jan. 27, 2016, https://www.theglobeandmail.com/news/politics/liberals-to -announce-new-transition-rules-for-assessing-pipelines/article28412555/ (accessed August 17, 2018).

6. Geoffrey Morgan, "National Energy Board doing 'inadequate' job of tracking whether pipelines meet approval," *Financial Post*, Jan. 26, 2016.

7. One is reminded of the impossible task set for nuclear engineers by the Finnish government in Michael Madsen's documentary film, *Into Eternity* (2010): to develop a system to ensure that no one will visit Onkalo, a storage site for radioactive waste, for at least 100,000 years.

8. Globe Editorial, "On Energy East, Trudeau has to be both referee and leader," *Globe and Mail*, Jan. 26, 2016, http://www.theglobeandmail.com/try-it -now/?articleId=28403383 (accessed August 17, 2018).

9. Globe Editorial, "Canada's greenhouse gas emissions can't be cut without a little pain," *Globe and Mail*, Feb. 4, 2016, http://www.theglobeandmail.com /try-it-now/?articleId=28560158 (accessed August 17, 2018).

10. Kyla Mandel, "Canada's Oil Exports Up 65 Per Cent Over Last Decade," *The Narwhal*, Feb. 22, 2016, https://thenarwhal.ca/canada-s-oil-exports-65-over-last -decade (accessed August 17, 2018).

11. Collis's *Once in Blockadia*, details the politics of the protests and ponders the power of poetry to unnerve the social imaginary shaped and supported by oil. See Collis, *Once in Blockadia*.

12. Geoffrey Morgan, "Burnaby calls on national energy regulator to suspend Trans Mountain pipeline review," *Vancouver Sun*, Jan. 20, 2016, http://www.vancouversun .com/news/burnaby+calls+national+energy+regulator+suspend+trans+mountain +pipeline+review/11664819/story.html (accessed August 17, 2018).

13. Justin Ling, "Canada Admits There's No Chance It'll Reach Its Climate Change Targets—Not Even Close," *Vice News Canada*, Feb. 1, 2016, https://news.vice.com /article/canada-admits-theres-no-chance-itll-reach-its-climate-change-targets -not-even-close (accessed August 17, 2018).

14. Barney, "Pipelines," in *Fueling Culture*, 269.

15. Natural Resources Canada, *Pipelines Across Canada*, para. 2, https://www.nrcan .gc.ca/energy/infrastructure/18856 (accessed August 17, 2018).

16. Barney, "Pipelines," 267.

17. Barney, "Pipelines," 268.

18. Writing from the perspective of an architect and critical geographer, Rania Ghosn generates a similar list of the multiple cultural, social, and political inscriptions of pipelines—all of which are rendered invisible by this apparatus once completed:

> The construction of such a large engineering project involved resolving labor availability, training, and expertise, as well as conditions of capital and technology. It meant deciding on the movement of local populations, on procurement of pipes and machinery, on whom to employ to construct and operate the pipeline, and how to secure it. Often operating in regions isolated from central power and unconnected to national and regional networks, the transport operation had to "develop" the frontier by deploying roads, ancillary services, and security posts. Simultaneously, the pipeline was built in public relations, in glossy brochures, colorful photos of communities and landscapes, and promises about positive impacts on people along the route. In its multiple dimensions, the fixation of the circulatory system in space produced a territory—simultaneously epistemological and material—through which international oil companies, transit and petro-states, and populations negotiated their political rationalities (Ghosn, "Territories of Oil: The Trans-Arabian Pipeline," 167–68).

19. Jones, *Routes of Power*, 124.

20. Jones, *Routes of Power*, 139.

21. Mitchell, *Carbon Democracy*, 36.

22. Mitchell, *Carbon Democracy*, 39.
23. Mitchell claims that from the 1920s "60 to 80 percent of the world oil production was exported." Mitchell, *Carbon Democracy*, 37.
24. Jones, *Routes of Power*, 143.
25. Ghosn, "Territories of Oil," 166.
26. Heidegger outlines the idea of "standing reserve" in "The Question Concerning Technology," in *The Question Concerning Technology and Other Essays*, 3–35.
27. Easterling, *Extrastatecraft*, 15.
28. Note: this is a very different idea of objects as actants than that promoted by Jane Bennett and others who want to give inanimate objects an efficacy. For Bennett, the fact that electricity, for example, can do things seemingly on its own, demands that we understand that materiality has a vitality to which we need to attend. What this vision of "vibrant matter" misses entirely is that the objects and infrastructures can be fashioned to be actants of a very particular kind—that is, to support human desires and actions, and so are even ontologically contained with the world of capital from which they emerge. See Bennett, *Vibrant Matter*.

    By "political invisibility," I do not mean to suggest that pipelines did not participate in state bureaucratic mechanisms of oversight and control. The "invisibility" I am speaking of here is to the space of political critique and contestation—that is, politics proper. I thank Arthur Mason for this point of clarification.
29. Rancière, "Comments and Responses."
30. Might this be a generational change? Perhaps; one of the reasons why the Harper government was so heavy-handed in their advocacy of Keystone XL was that they simply didn't grasp what was at issue: for them, a pipeline belongs in the space of extrastatecraft, something whose obvious structural and technological necessity meant that it didn't require the intervention of governments or necessitate political decision-making of any kind; for many others—including those who voted for the Trudeau government—extrastatecraft is a matter to which states are forced to attend, for ecological reasons if no other.
31. The 1995 decision by the Government of Canada to recognize the inherent right of First Nations to self-government (under the terms set out in Section 35 of the Constitution Act of 1982) has been one of the reasons that they have begun in recent decades to more strongly assert their rights as sovereign communities.
32. Aaron Wherry, "Accused of 'swanning' around Davos, Trudeau called to mediate Energy East," *CBC News*, Jan. 25, 2016, http://www.cbc.ca/news/politics/accused-of-swanning-around-davos-trudeau-called-to-mediate-energy-east-1.3419179 (accessed August 17, 2018).
33. John Paul Tasker, "Trudeau, Coderre meet after Tories blast Energy East comments," *CBC News*, Jan. 25, 2016, http://www.cbc.ca/news/politics/ambrose-energy-east-national-unity-crisis-1.3418664 (accessed August 17, 2018).
34. See, for instance, Steven Chase, "Trudeau's policies divide Canada, Manning asserts," *Globe and Mail*, Feb. 27, 2016, A4. The former leader of Canada's Reform Party, Preston Manning, invoked national unity in relation to pipeline projects at an annual meeting of the Canadian conservative movement: "What will be the unity consequences when a supposedly national government welcomes tankers

bringing in foreign oil on the east coast but wants to ban tankers on the west coast from carrying Canadian oil to world markets?" he said in remarks critical of the lack of progress on new pipeline capacity.

35. Barry, *Material Politics*, 12.
36. Szeman, "On Energopolitics."
37. See, for example, Éric Darier, "Foucault and the Environment: An Introduction."
38. Knox, "Footprints in the City," 415.
39. Knox, "Footprints in the City," 415.
40. Dipesh Chakrabarty makes a similar point. "There is no politics that corresponds to planetary perspectives," he writes. "Humans face the emerging phenomenon of planetary warming from a default position, that is, from within the politics of the institutions that were created to deal with the 'globe' of 'globalization' with all the assumptions of 'stable' Holocene conditions built into them." See Chakrabarty, "Afterword," 168.
41. Caffentzis, "The Petroleum Commons."
42. Chakrabarty, "Afterword" and "The Politics of Climate Change Is More Than the Politics of Capitalism."

CHAPTER 12

———

On the Politics of Region (2018)

I was invited to contribute to "Dimensions of Citizenship," an editorial collaboration between the journal *e-flux Architecture* and the US Pavilion at the 2018 Venice Architectural Biennale. "Dimensions of Citizenship" was organized around seven commissioned works, each exploring a different scale of belonging. The overall aim of this collaboration was to critically examine the spatial constructions of citizenship, its inclusions and exclusions, at different and distinct scales of belonging. I was asked to write a text on the scale of "region," which was being explored at the Biennale by the New York-based studio SCAPE. They described the impulse for their interest in region in the following terms:

> The U.S. regional belts—corn belt, rust belt, Bible belt, sun belt—are rich parts of the American identity. Typically organized around an area's shared resources—oil, agriculture, steel production or fertile soil—regions are also defined by shared cultural values. Yet as politics, the global marketplace, and climate change lead to factory closures, pipeline fights, automation, or massive flooding and drought, how might designers respond to frayed notions of citizenship? Poised against the backdrop of profound environmental and economic change in the U.S., works exhibited at this scale argue that architecture and landscape architecture can play a role in transforming and redefining regions.

My essay uses SCAPE's investigation as a way of expanding the critical and political possibilities introduced by the concept of region—a geographic scale that challenges the self-certainties of the local and the national, of typical modes of belonging and community, and the place of the environment in our understanding of citizenship.

Indifferent Systems

In November 2017, the Nebraska Public Service Commission voted 3–2 in favor of allowing the proposed Keystone XL pipeline to be built in the state. Actively opposed by groups such as 350.org and blocked during the presidency of Barack Obama, Keystone XL was revived by the Trump administration as one of its very first acts of state. When finished, it will travel over 1000 miles (1600 km) from Hardisty, Alberta, to Steele City, Nebraska, where it will join up with the existing Keystone network. The pipeline system will move Canadian oil from the Alberta oil sands and US oil from Montana and North Dakota to refineries on the Texas Gulf Coast and to tank farms in Cushing, Oklahoma, and Patoka, Illinois. It will have a capacity of up to 1.1 million barrels per day. For a world still hungry for fossil fuels, this "export limited" (the "XL" in the name) pipeline is expected to generate profits for TransCanada Corporation—the owners of Keystone—as well as the oil companies that use its massive infrastructure to move their products to market.

One of the surprises of the Nebraska decision was that the Commission shifted the pipeline's route through the state. Instead of sticking with TransCanada's preferred route, for which the company had already secured easements from the majority of impacted property owners, the Commission proposed to follow an existing Keystone line through the area. Environmentalists who object to the expansion of the fossil-fuel system and worry about the possible repercussions of oil spills along pipeline routes—like the one TransCanada was dealing with in South Dakota at the time of the decision—drew attention to the fact that no environmental studies had been done on this new route. Impacted landowners objected for similar reasons, as well as for the fact that no easements had been negotiated. While TransCanada expressed unhappiness with the extra cost implications of the new route, they were also pleased that, at long last, a decision had been made and the final stage of the full Keystone system could be built.

Pipelines like the Keystone XL obliterate the spaces and environments that exist between an oil source and its end users. They cut straight lines across landscapes, indifferent to the specifics of geography. For city dwellers, these technologies for transporting energy generate indifference not only of a spatial kind, but also of an ethical or political one: extraction zones and networks of transport have little impact on the majority of those who use fossil fuels, for whom the stuff of energy appears, as if by magic, in their furnaces or at gas stations. Yet the protracted and public struggle over Keystone XL shows that other lines on maps—the borders of states or countries, the lines around property—can inhibit or block the easy passage of pipelines through space. Property owners can, at a minimum, make financial claims against pipeline companies when ribbons of steel cut across their fields and gardens. Everything else that shapes geography—from distinctive geologies, watersheds, and animal habitats to indigenous communities and histories of human habitation—gets ignored and left out of cost-benefit calculations.

From the perspective of the Nebraska Commission, which nudged the vector of the pipeline without considering the regions it would endanger, one route is as good as any other. Yet regions are where the consequences of technologies—whether physical technologies such as pipelines or the *technē* of governments that establish borders and property—are felt most determinately. The region in which the November 2017 Keystone spill of 5000 barrels took place will take years to clean up (a 2016 spill of 400 barrels at another site in South Dakota took ten months to ameliorate).[1] Spills always take place in-between, in the space of the region, as far away from the abstract legislative space of a state or country as the cities to which the black pools of fuel were intended to move.

Region is a concept that we rarely consider, even if much of the trauma and crisis of modernity is happening there. We need to understand the dynamics of region if we are going to challenge the indifferent systems of infrastructure and politics that carve up and control everything in their orbit. Region owes nothing to the forms of citizenship granted to subjects by states, or to the power that comes from indifference to the spaces traversed by the infrastructure of modernity. Might region allow us to understand anew the connections of space, belonging, and environment needed to take on the political challenges of our era?

Toward a Theory of Region

On a scale from the global to the local, the region hovers somewhere in-between. *Where* in-between is difficult to determine, in part because region lacks any precise definition. A region can be an expanded sense of the local—a city and its suburban and exurban pseudopods. It can be comprised of a series of nations, linked by trade, history, religion, or ethnicity. Region can point to zones within nations, demarcated by as little as the points on a compass (Northeast, West, Midwest, etc.). It can be understood in relation to religion (Bible Belt, Borscht Belt, Jell-O Belt), or be shaped around labor and industry (Wheat Belt, Rust Belt). Region can also be defined in relation to geology (the Rockies, the Mississippi Delta), a configuration that overlaps inexactly with the spaces staked by existing political forms and laws that make a claim on spaces and their inhabitants (e.g., the Rockies run through both Canada and the United States).

Region is thus a messy term. But to develop a precise definition of region would be to miss the point of the demand that the concept makes on us. It is in the indistinct nature of region, in the broad and shifting set of characteristics and qualities that extend from one idea of region (the Middle East) to another (the Ogallala Aquifer), that its power lies.

Inherent in every idea of region is a contiguity of geographic space. On a map, regions are paramecium-like zones of connection that refuse to obey the sharp lines of state borders or property. No nation-state is a region (even if region is sometimes used to designate groups of nations), and no region has laws, police, and military—the apparatus of modern power that has so deeply shaped ideas of community and subjectivity. Regions rub raw the self-certainties of modern state formations. A region is that spot on the Achilles tendon (of capitalism, property, or liberal democratic governance) where an ill-fitting shoe raises a bloody patch and threatens to sever altogether both the tendon and the power mythically contained within it.

In addition to contiguity of space, what is of primary importance to a region is that there is a *there*. Every region can be seen as a type of ecology—an environment (a contiguous geographic zone), the subjects that animate it (whether these are animals and plants, specific religious groups, the resources that lie beneath the ground, or the strata of the inanimate), and the relation between these two. Just as important to note is that there is never a single *there* there, but rather, of necessity, a rich, heterogeneous set of overlapping ecologies that speak to the multiple relations that exist in any geographic zone.

Against the abstraction of the nation-state and other political boundaries, regions demand that we be alert to the innumerable ecologies that constitute the lives of individuals and their communities. Their multiplicity asks not that we try to name all of these relations in order to codify them into some new logic of regionality. The necessary multiplicity of ecologies, each environment linked in an essential way to the organisms that dwell there (people, animals, plants, fuel, minerals, non-humans, forces, processes), asks that we undo the abstract mechanisms of power, which pay little attention to the planet's ecologies. These mechanisms of power operate instead via well-established modes of power linked to inclusion and exclusion. Nations and cities do not seem to pay attention to the demands that multiple ecologies make on them. Regions, on the other hand, are deeply attuned to the realities of the shifting ideas and realities of being there—including the *there* of nations and cities—and spill over and beyond all established political borders.

Regions thus pose a challenge to ideas of citizenship that lie at the heart (of the myth) of the modern democratic state. A citizen belongs to a nation-state. With this belonging come responsibilities (such as defending the nation, when necessary) as well as, at least potentially, rights and opportunities (security from internal and external threats, education and health, and the right to own property). One of the powers of the concept and practice of citizenship is that it insists that all citizens within the spaces it delineates are equal, i.e., there are no regional differences and degrees of citizenships. A map of the globe shows every nation-state to be a single, flattened color. The jigsaw shapes of red, blue, green, and yellow differentiate the citizens of one country from another, and at the same time assert that inside these flattened color fields all citizens are the same. But in truth, citizenship is shaped around inclusions and exclusions, around violent delineations of belonging. The differential quality of citizenship extends not just to who one is, but what one does and where one lives within a nation-state. Even before the law, there are enormous inequalities between citizens, which are recognized by the disparate zones of policing, incarceration, and state violence against its citizens. There are zones of opportunity and hardship, wealth and poverty, black and white. Regions speak the lie of nations with respect to their claim on space by drawing attention to differential experiences of citizenship that exist within any nation at any point in its existence.

There are still other regions to which we need to be alert—regions of dispossession, of poverty, and wealth—which emerge from the extension of the logics of capitalism to the whole of the planet. Free trade zones, spaces of cheap labor, abandoned spaces of productions—all emptied of the faintest traces of workers' rights—appear as spaces in which nations have forsaken the commitments that they might have once promised to their citizens in favor of the logics of globalization that would render the entire world (and not just the internal space of nations) into a single indifferent space in which the logic of capitalism rules.

The environmental consequences of global, neoliberal capitalism create further regions—of eco-destitution, monocultures, commodity frontiers, soils drained of life, polluted geographies. These toxic ecologies emerge out of the utter indifference shown by capitalist states to respecting, preserving, or honoring the specificity and complexity of regions. These spaces might often be described in the language of region (as in: "regions of pollution"), but they are regions in name only. These ecologies are the consequence of the abstract logic of power, control, property, and profit, the outcome of the instrumentality of statecraft as well as extrastatecraft (the protocols and standards that shape modern professional activity in the design and control of structures and infrastructures).[2]

It is essential to grapple with the forces that generate these toxic ecologies, since they offer up a false idea of political change. One can understand why inhabitants of dispossessed and distressed regions might want to restore what was removed or taken away, such as the factories that once provided jobs. But it is a mistake to imagine this dispossession as a deficit of true citizenship that now needs to somehow be restored or made full (the national restored from the global; regional differences inside the nation flattened out to restore the promised equality of citizenship). Neoliberalism has ensured that *all* citizenship has been torn asunder; it no longer even hides behind the narrative that citizenship ensures equality, as it might once have done. If citizenship is based on inclusion and exclusion, then it is only natural that those once included in its fold should want the safety and privilege that had been (minimally) accorded them. Yet at a moment when we need to pay greater attention to the complex ecologies we inhabit, citizenship is a crude, abstract device of being and belonging that obscures the multiplicity and heterogeneity of regions and makes it impossible to address the forces that generate the toxicity of all too many regions on the planet.[3]

We have to be alert to the limits, too, of existing practices and techniques of region, such as those that exist in the practice of architecture. "Critical regionalism" names structures that employ the codes and character of modern architecture, while also paying attention to a building's geographic context.[4] Even as it challenges the flattening universality of dominant forms of architecture (such as the International Style), critical regionalism foregrounds geography in a way that provides little more than shading to a style—the modern—that exists before and beyond it. At best, the region in critical regionalism appears as small adjustments that different geographies might demand of modern architecture. These modern buildings can congratulate themselves for being attuned to differences of landscape, while still being modern and not truly dealing with the challenge of region at all.

Critical regionalism speaks only to a single ecology—the environment of modern buildings, which exist within the dictates and demands of capitalism. A truly critical practice of region would instead explore multiple ecologies, attending to the full range of relationships that exist in any geography. What defines the environment and organisms that make up a region can and should vary widely. How might architecture respond to regions defined, for instance, by different forms of energy? Or by diverse modes of labor and income? Or by immigration and political counter-narratives? How might an architectural practice react to and interact with multiple regions—with geographies of living and non-living bodies, or with geological and environmental zones that might assert that the construction of modern architecture is akin to the abstraction of the nation-state, whatever shading of style one might apply?

To the Region!

The Keystone XL pipeline system operates both in and outside of the logic of region. It draws on discrete regions of resources (e.g., the

Athabasca Tar Sands, the Bakken Oil Shale), artificially assembling them together in order to move them elsewhere. Region matters to the XL, but only as sites of entry and exit. Everything in-between is a hindrance, whether due to scale (the length of pipeline requires forty-one pump stations to keep the oil moving), the geology it has to navigate (e.g., the Missouri River, the Sandhills of Nebraska, or the Ogallala aquifer, which stretches from South Dakota to Texas), or its impact on communities that might challenge its logics and imbedded presumptions. While the Keystone XL and pipelines like it depend on the existence of resource regions, the primary logic they follow is that of capitalist modernity and its understanding of resource as a "standing reserve."[5]

The flattening of geography, geology, and community (and every other region) along the path of a pipeline mirrors the operations of citizenship, a political infrastructure that uses the cloak of equality (especially before the law) to deny or disavow all manner of inequalities and the disequilibria between its citizens. The logic of region laid out here is intended as a heuristic rather than something like a law of nature—rough around the edges, not always workable in each and every evocation of region. Attention to region offers a rejoinder to protocols of dividing up space that do not attend to the rich and multiple ecologies that exist there. As the path of a pipeline shows us, the sovereign space of the state and private property, along with other practices of spatializing, have consequences for the ecologies of these regions. Yet what a pipeline can never name are the relations between people, place, environment, objects, animals, gases, and plants that inhabit and shape a region. If for no other reason than the environmental challenges we face, we need now to actively and attentively inhabit non-flattened, non-empty spaces.

Could region act as a possible site of citizenship—an alternative to the one to which we presently seem fated? We have examples of regions creating powerful new forms of belonging and community. The protests that took place near the Standing Rock Indian Reservation against the Dakota Access Pipeline in 2016 and 2017 brought together Indigenous people from across North America, as well as other protestors. Those assembled at Standing Rock wanted to draw attention to the threat of the pipeline to sacred burial grounds as well as to the quality of the community's water. Attempts by members of the Standing Rock Sioux Tribe to use the mechanisms of official citizenship (e.g., appeals to the Advisory Council on Historic Preservation, the Department of Interior, the Environmental Protection Agency, and suits filed in court) went nowhere. And so a new community—one intimately alert to the reality of region and the multiplicity of ecologies in their lifeworlds, one made up of Sioux and those opposed to the pipeline from around the world—came into being in the protest camps that blocked the abstract, indifferent logics of the pipeline.

It seems difficult to adapt the old language of citizenship to truly new modes of being and belonging in space, like those enacted at Standing Rock. The moment one declares citizenship in relation to a region, it fixes a single ecology in place, creating something akin to a micro-nation. Citizenship is a damaged concept that insists on inclusions and exclusions, on the establishment of borders and sovereignty. It might be that region allows us rethink how to commit to one another and to the ecologies we inhabit without the necessity of sovereignty. Region constitutes a powerful redefinition of the political on a planet where borders have threatened the health of communities and ecologies far more than it has helped them.

NOTES

1. See Mitch Smith and Julie Bosmannov, "Keystone Pipeline Leaks 210,000 Gallons of Oil in South Dakota," *New York Times*, Nov. 16, 2017, https://www.nytimes .com/2017/11/16/us/keystone-pipeline-leaks-south-dakota.html?_r=0 (accessed August 17, 2018).
2. The term "extrastatecraft" was coined by Keller Easterling in *Extrastatecraft: The Power of Infrastructure* to describe protocols and standards connected to the design of structures and infrastructures, which generate outcomes that support and amplify neoliberalism.
3. For an overview of the problems and limits of a view of neoliberalism as an enclosure of the commons (that needs to be rectified by restoring some of the powers of the state), see Dardot and Laval, *Common: On Revolution in the 21st Century*, especially Part 2.
4. The idea of "critical regionalism" is elaborated in Frampton, "Towards a Critical Regionalism."
5. Martin Heidegger writes that over the course of modernity, "Nature becomes a gigantic *gasoline* station, an energy source for modern technology and industry." Heidegger, *Discourse on Thinking*, 50.

References

Adorno, Theodor. *Aesthetic Theory*. Translated by Robert Hullot-Kentor. Minneapolis: University of Minnesota Press, 1998.

———. "On the Question: 'What is German?'" In *Critical Models: Interventions and Catchwords*. Translated by Henry W. Pickford. New York: Columbia University Press, 1998.

———. *Minima Moralia: Reflections from Damaged Life*. Translated by E. F. N. Jephcott. London: New Left Books, 1974.

———. "Theses on the Sociology of Art." Translated by Brian Trench. *Working Papers in Cultural Studies* 2 (1972): 121–28.

Agamben, Giorgio. *Homo Sacer: Sovereign Power and Bare Life*. Translated by Daniel Heller-Roazen. Stanford: Stanford University Press, 1998.

Ahmad, Aijaz. *In Theory: Class, Nations, Literatures*. New York: Verso, 1992.

Althusser, Louis. *On the Reproduction of Capitalism*. Translated by G. M. Goshgarian. New York: Verso, 2014.

Anderson, Perry. *The Origins of Postmodernity*. New York: Verso, 1998.

Appadurai, Arjun. *Modernity at Large: Cultural Dimensions of Globalization*. Public Worlds. Vol. 1. Minneapolis: University of Minnesota Press, 1996.

Arnold, Matthew. *Culture and Anarchy*. Oxford: Oxford University Press, 2006.

Arns, Inke, ed. *The Oil Show*. Berlin: Revolver, 2011.

Arrighi, Giovanni. "Hegemony Unravelling—I." *New Left Review* 32 (2005): 23–80.

Ashcroft, Bill, Gareth Griffiths, and Helen Tiffen. *Key Concepts in Postcolonial Studies*. New York: Routledge, 1998.

Balibar, Étienne, and Pierre Macherey. "On Literature as an Ideological Form: Some Marxist Propositions." In *Untying the Text: A Post-Structuralist Reader*. Edited by Robert Young. London: Routledge & Kegan Paul, 1981, 79–99.

Barney, Darin. "Pipelines." In *Fueling Culture: 101 Words on Energy and Environment*. Edited by Imre Szeman, Jennifer Wenzel, and Patricia Yaeger. New York: Fordham University Press, 2017, 267–70.

Barrett, Ross, and Daniel Worden, eds. *Oil Culture*. Minneapolis: University of Minnesota Press, 2014.

Barry, Andrew. *Material Politics: Disputes Along the Pipeline*. New York: Wiley-Blackwell, 2013.

Barthes, Roland. *Mythologies*. Translated by Annette Lavers. New York: Noonday Press, 1972.

Baucom, Ian. "History 4°: Postcolonial Method and Anthropocene Time." *Cambridge Journal of Postcolonial Literary Inquiry* 1, no. 1 (March 2014): 123–42.

Bauman, Zygmunt. *Globalization: The Human Consequences*. Cambridge, UK: Polity, 1998.

———. *Wasted Lives: Modernity and Its Outcasts*. Cambridge, UK: Polity, 2004.

Beck, Ulrich. *Risk Society: Towards a New Modernity*. London: Sage, 1992.

Beckett, Samuel. *Company; Ill Seen Ill Said; Worstward Ho; Stirrings Still.* Edited by Dirk van Hulle. London: Faber, 2009.

Bellamy, Brent Ryan, and Jeff Diamanti, eds. *Materialism and the Critique of Energy.* Chicago: MCM Prime, 2018.

Bellamy Foster, John. "The Ecology of Marxian Political Economy." *Monthly Review* 63, no. 4 (2011). http://monthlyreview.org/2011/09/01/the-ecology-of-marxian-political-economy. Accessed August 17, 2018.

Benjamin, Walter. "The Work of Art in the Age of Mechanical Reproduction." In *Illuminations: Essays and Reflections.* Translated by Harry Zohn. New York: Schocken Books, 1968.

Bennett, Jane. *Vibrant Matter: A Political Ecology of Things.* Durham, NC: Duke University Press, 2010.

Berger, Thomas R. *Northern Frontier, Northern Homeland: The Report of the Mackenzie Valley Pipeline Inquiry.* Revised edition. Vancouver: Douglas & McIntyre, 1988.

Berlant, Lauren. "Cruel Optimism." *differences: A Journal of Feminist Cultural Studies* 17, no. 3 (2006): 20–36.

Berlinger, Joe, dir. *Crude: The Real Price of Oil.* 2009; New York: First Run Features, 2010. DVD.

Bernal, Martin. *Black Athena: The Afroasiatic Roots of Classical Civilization.* New Brunswick, NJ: Rutgers University Press, 1989.

Bhabha, Homi K. "Signs Taken for Wonders." *Critical Inquiry* 12, no. 1 (1985): 144–65.

Biemann, Ursula, dir. *Black Sea Files: Video Essay in 10 Parts.* 2005; Video Databank.

Biro, Andrew. *Denaturalizing Ecological Politics: Alienation from Nature from Rousseau to the Frankfurt School and Beyond.* Toronto: University of Toronto Press, 2005.

Bonneuil, Christophe, and Jean-Baptist Fressoz. *The Shock of the Anthropocene.* Translated by David Fernbach. New York: Verso, 2016.

Borasi, Giovanna, and Mirko Zardani. *Sorry, Out of Gas: Architecture's Response to the 1973 Oil Crisis.* Montreal: Canadian Centre for Architecture/Corraini Edizioni, 2008.

Bourdieu, Pierre. *Acts of Resistance: Against the Tyranny of the Market.* Translated by Richard Nice. New York: The New Press, 1998.

———. *Distinction: A Social Critique of the Judgement of Taste.* Translated by Richard Nice. Cambridge, MA: Harvard University Press, 1984.

———. *The Rules of Art: Genesis and Structure of the Literary Field.* Translated by Susan Emanuel. Stanford, CA: Stanford University Press, 1996.

Boyer, Dominic. "Energopolitics and the Anthropology of Energy." *Anthropology News* 52, no. 5 (May 5, 2011): 5–7.

Bozak, Nadia. *The Cinematic Footprint: Lights, Camera, Natural Resources.* New Brunswick, NJ: Rutgers University Press, 2011.

Brecht, Bertolt. *Brecht on Theatre.* Edited and translated by John Willett. New York: Hill and Wang, 1977.

Brennan, Timothy. "Cosmo-Theory." *The South Atlantic Quarterly* 100, no. 3 (2001): 659–91.

Brogan, Chris. *The Freaks Shall Inherit the Earth: Entrepreneurship for Weirdos, Misfits, and World Dominators.* Hoboken, NJ: Wiley, 2014.

Brooks, David. *Bobos in Paradise: The New Upper Class and How They Got There*. New York: Simon and Schuster, 2001.

Brown, Nicholas. *Utopian Generations: The Political Horizon of Twentieth-Century Literature*. Princeton, NJ: Princeton University Press, 2005.

Brown, Nicholas, and Imre Szeman. "What is the Multitude? Questions for Michael Hardt and Antonio Negri." *Cultural Studies* 19, no. 3 (2005): 372–87.

Buck-Morss, Susan. *Dreamworld and Catastrophe*. New York: MIT Press, 2001.

Budgen, Sebastian. "A New 'Spirit of Capitalism.'" *New Left Review* 1 (January–February 2000): 149–56.

Buell, Frederick. "A Short History of Oil Cultures: Or, the Marriage of Catastrophe and Exuberance." *Journal of American Studies* 46, no. 2 (2012): 273–93.

———. "Nationalist Postnationalism: Globalist Discourse in Contemporary American Culture." *American Quarterly* 50, no. 3 (1998): 548–91.

Bull, Malcolm. "Between the Cultures of Capital." *New Left Review* 11 (2001): 95–114.

Bürger, Peter. *Theory of the Avant-Garde*. Translated by Michael Shaw. Minneapolis: University of Minnesota Press, 1985.

Burkett, Paul, and John Bellamy Foster. "Metabolism, Energy, and Entropy in Marx's Critique of Political Economy." *Theory and Society* 35, no. 1 (2006): 109–56.

Burtless, Gary, et al. *Globaphobia: Confronting Fears about Open Trade*. Washington, DC: Brookings Institution Press, 1998.

Burtynsky, Edward. *Oil*. London: Steidl, 2011.

Butler, Judith. *Precarious Life: The Powers of Mourning and Violence*. London: Verso, 2006.

Caffentzis, George. "The Petroleum Commons." *Counterpunch*, Dec. 15, 2004. www.counterpunch.org/2004/12/15/the-petroleum-commons/. Accessed August 17, 2018.

Callison, Candis. *How Climate Change Comes to Matter*. Durham, NC: Duke University Press, 2014.

Canadian Association of Petroleum Producers. *Upstream Dialogue: The Facts on Oil Sands*. Calgary: CAPP, April 2012.

Cantillon, Richard. *An Essay on Economic Theory: An English Translation of Richard Cantillon's Essai sur la nature du commerce en général*. Edited by Mark Thornton. Translated by Chantal Saucier. Auburn, AL: Ludwig von Mises Institute, 2010.

Čapek, Karel. *The Absolute at Large*. London: Allen, 1944.

Casanova, Pascale. *The World Republic of Letters*. Cambridge, MA: Harvard University Press, 2005.

Cazdyn, Eric, and Imre Szeman. *After Globalization*. London: Wiley-Blackwell, 2011.

Cenovus. "Canadian Ideas at Work" (advertisement). YouTube, 2011. www.youtube.com/watch?v=j0vYTFve7tA. Accessed August 17, 2018.

Chakrabarty, Dipesh. "Afterword." *The South Atlantic Quarterly* 116, no. 1 (2017): 163–68.

———. "The Climate of History: Four Theses." *Critical Inquiry* 35, no. 2 (Winter 2009): 197–222.

———. "The Politics of Climate Change Is More Than the Politics of Capitalism." *Theory, Culture, and Society* 34, no. 2-3 (2017): 25-37.

———. *Provincializing Europe: Postcolonial Thought and Historical Difference*. Princeton, NJ: Princeton University Press, 2000.

Chambers, Iain. *Culture after Humanism.* New York: Routledge, 2001.

Charlesworth, J. J. "Any Other But Our Selves." *Mute: Culture and Politics after the Net.* September 25, 2008. http://www.metamute.org/editorial/articles/any-other -our-selves. Accessed August 17, 2018.

Chesky, Brian. "Shared City." *Medium.* March 26, 2014. www.medium.com/@bchesky /shared-city-db9746750a3a. Accessed August 17, 2018.

Ching, Leo. "Globalizing the Regional, Regionalizing the Global: Mass Culture and Asianism in the Age of Capital." *Public Culture* 12, no. 1 (2000): 233–57.

Chua, Amy. *Day of Empire: How Hyperpowers Rise to Global Dominance—and Why They Fall.* New York: Doubleday, 2007.

Clausen, Christopher. "'National Literatures' in English: Toward a New Paradigm." *New Literary History* 25, no. 1 (1994): 61–72.

———. "Nostalgia, Freedom, and the End of Cultures." *Queen's Quarterly* 106, no. 2 (1999): 233–44.

Colás, Santiago. "The Third World in Jameson's *Postmodernism or the Cultural Logic of Late Capitalism.*" *Social Text* 31–32 (1992): 258–70.

Colebrook, Claire. "Not Symbiosis, Not Now: Why Anthropogenic Climate Change Is Not Really Human." *The Oxford Literary Review* 34, no. 2 (2012): 185–210.

Coll, Steve. *Private Empire: ExxonMobil and American Power.* New York: Penguin, 2012.

Collis, Stephen. *Once in Blockadia.* Vancouver: Talonbooks, 2016.

Comaroff, Jean, and John L. Comaroff, "Millennial Capitalism: First Thoughts on a Second Coming." *Public Culture* 12, no. 2 (2000): 291–343.

Constanza, Robert et al. "The Value of the World's Ecosystem Services and Natural Capital." *Nature* 387 (1997): 253–60.

Cornwall, Andrea, and Karen Brock. "What Do Buzzwords Do for Development Policy? A Critical Look at 'Participation,' 'Empowerment' and 'Poverty Reduction.'" *Third World Quarterly* 26, no. 7 (2005): 1043–60.

Cultural Industries Sectoral Advisory Group on International Trade (Canada). *Canadian Culture in a Global World: New Strategies for Culture and Trade.* Ottawa, 1999.

Darby, Phillip. *The Fiction of Imperialism: Reading between International Relations and Postcolonialism.* London: Continuum, 1998.

Dardot, Pierre, and Christian Laval. *Common: On Revolution in the 21st Century.* Translated by Michael MacLellan. London: Bloomsbury, 2018.

———. *The New Way of the World: On Neoliberal Society.* New York: Verso, 2014.

———. "The New Way of the World, Part II: The Performance/Pleasure Apparatus." *e-flux,* no. 52 (2014). http://www.e-flux.com/journal/the-new-way-of-the-world- part-ii-the-performancepleasure-apparatus/. Accessed August 17, 2018.

Darier, Éric. "Foucault and the Environment: An Introduction." In *Discourses of the Environment.* Edited by Éric Darier. Malden, MA: Blackwell, 1999, 1–33.

Darley, Julian. *High Noon for Natural Gas.* White River Junction, VT: Chelsea Green, 2004.

Davis, Mike. *Planet of Slums: Urban Involution and the Informal Working Class.* London: Verso, 2006.

Debeir, Jean-Claude, Jean-Paul Deléage, and Daniel Hémery. *In the Servitude of Power: Energy and Civilization through the Ages*. Translated by John Barzman. London: Zed Books, 1991.

Deffeyes, Kenneth. *Beyond Oil: The View from Hubbert's Peak*. New York: Hill and Wang, 2005.

———. *Hubbert's Peak: The Impending World Oil Shortage*. Princeton, NJ: Princeton University Press, 2001.

Diamanti, Jeff. "Aesthetic Economies of Growth: Energy, Value, and the Work of Culture After Oil." PhD Thesis, University of Alberta, 2015.

———. "Energyscapes, Architecture, and the Expanded Field of Postindustrial Philosophy." *Postmodern Culture* 26, no. 2 (2016), n.p.

———. "Three Theses on Energy and Capital." *Reviews in Cultural Theory* 6, no. 3 (2016): 13–16.

Dorfman, Ariel, and Armand Mattelart. *How to Read Donald Duck*. New York: International General, 1975.

Eagleton, Terry. *The Idea of Culture*. Oxford: Blackwell, 2001.

———. *The Ideology of the Aesthetic*. Malden, MA: Blackwell, 1990.

Easterling, Keller. *Extrastatecraft: The Power of Infrastructure Space*. New York: Verso, 2014.

Eggers, Dave. *The Circle*. New York: Knopf, 2013.

Elkins, James. *Pictures and Tears: People Who Have Cried in Front of Paintings*. New York: Routledge, 2001.

Elliott, Jane. "Suffering Agency: Imagining Neoliberal Personhood in North America and Britain." *Social Text* 115, no. 2 (2013): 83–101.

Fabian, Johannes. *Time and the Other: How Anthropology Makes Its Object*. New York: Columbia University Press, 1983.

Federici, Silvia. "From Commoning to Debt: Financialization, Microcredit, and the Changing Architecture of Capital Accumulation." *The South Atlantic Quarterly* 113, no. 2 (2014): 231–44.

Fisher, Mark. *Capitalist Realism: Is There No Alternative?* Winchester, UK: Zero Books, 2009.

Fisker, Jacob Lund. "The Laws of Energy." In *The Final Energy Crisis*. Edited by Andrew McKillop and Sheila Newman. London: Pluto Press, 2005, 74–86.

Florida, Richard. *Cities and the Creative Class*. New York: Routledge, 2004.

———. *The Flight of the Creative Class*. New York: Harper Business, 2005.

———. *The Great Reset: How New Ways of Living and Working Drive Post-Crash Prosperity*. Toronto: Random House, 2010.

———. "How the Crash Will Reshape America." *Atlantic Monthly*, March 2009.

———. *The Rise of the Creative Class*. New York: Basic Books, 2003.

———. *Who's Your City?: How the Creative Economy Is Making Where to Live the Most Important Decision of Your Life*. New York: Basic Books, 2008.

———. "The World is Spiky." *Atlantic Monthly*, October 2005: 48–51.

Frampton, Kenneth. "Towards a Critical Regionalism: Six Points for an Architecture of Resistance." In *Anti-Aesthetic: Essays on Postmodern Culture*. Seattle: Bay Press, 1983, 16–30.

Franco, Jean. "The Nation as Imagined Community." In *Dangerous Liaisons: Gender, Nations, and Postcolonial Perspectives*. Edited by Anne McClintock, Aamir Mufti, and Ella Shohat. Minneapolis: University of Minnesota Press, 1997, 130–37.

Fraser, Jill Andresky. *White Collar Sweatshop*. New York: W. W. Norton, 2002.

Fukuyama, Francis. *The End of History and the Last Man*. Toronto: HarperCollins Canada, 1993.

Gaines, Jane M. "Political Mimesis." In *Collecting Visible Evidence*. Edited by Jane M. Gaines and Michael Renov. Minneapolis: University of Minnesota Press, 1999, 84–102.

Gansky, Lisa. *The Mesh: Why the Future of Business Is Sharing*. New York: Penguin, 2010.

Garcia Canclini, Néstor. *Consumers and Citizens: Globalization and Multicultural Conflicts*. Translated by George Yúdice. Minneapolis: University of Minnesota Press, 2001.

Gelpke, Basil, and Ray McCormack, dirs. *A Crude Awakening: The Oil Crash*. 2006; Zurich, Lava Productions AG, 2007. DVD.

German Advisory Council on Global Change. *World in Transition 3: Towards Sustainable Energy Systems V. 3*. London: Earthscan, 2004.

German Federal Ministry for Economic Affairs. *The Energy of the Future: Fourth 'Energy Transition' Monitoring Report*. Berlin: Federal Ministry for Economic Affairs, 2015.

Ghosh, Amitav. *The Great Derangement: Climate Change and the Unthinkable*. Chicago: University of Chicago Press, 2016.

———. "Petrofiction: The Oil Encounter and the Novel." In *Incendiary Circumstances: A Chronicle of the Turmoil of Our Times*. New York: Houghton Mifflin Harcourt, 2007, 138–51.

Ghosn, Rania. "Territories of Oil: The Trans-Arabian Pipeline." In *The Arab City: Architecture and Representation*. Edited by Amale Andraos and Nora Akawi. New York: Columbia Books on Architecture and the City, 2016, 165–75.

Gilligan, Melanie, and Marina Vishmidt. "'The Property-less Sensorium': Following the Subject in Crisis Times." *The South Atlantic Quarterly* 114, no. 3 (2015): 611–30.

Golden, Mark. "State-owned Oil Companies Increase Price Volatility and Pollution, but Rarely Get Used as Geopolitical Weapons, Says Stanford Researcher." *Stanford Report*, Feb. 15, 2012. http://news.stanford.edu/news/2012/february/state-owned -oil-021512.html. Accessed August 17, 2018.

Goodstein, David. *Out of Gas: The End of the Age of Oil*. New York: W. W. Norton, 2005.

Graeber, David. "On the Phenomenon of Bullshit Jobs." *Strike! Magazine*, August 17, 2013. www.strikemag.org/bullshit-jobs/. Accessed August 17, 2018.

Gramsci, Antonio. *Selections from the Prison Notebooks*. Edited and translated by Quintin Hoare and Geoffrey Nowell Smith. New York: International Publishers, 1971.

Greenblatt, Stephen. "Racial Memory and Literary History." *PMLA* 116, no. 1 (2001): 48–63.

Grossberg, Lawrence. "Speculations and Articulations of Globalization." *Polygraph* 11 (1999): 11–48.

Habermas, Jürgen. *The Structural Transformation of the Public Sphere: An Inquiry into a Category of Bourgeois Society.* Translated by Thomas Burger and Frederick Lawrence. Cambridge, MA: MIT Press, 1991.

Habila, Helon. *Oil on Water: A Novel.* New York: W. W. Norton, 2011.

Hall, Stuart, and Alan O'Shea. "Common-sense Neoliberalism." In *After Neoliberalism? The Kilburn Manifesto.* Edited by Stuart Hall, Doreen Massey, and Michael Rustin. London: Lawrence and Wishart, 2013. www.lwbooks.co.uk/journals/soundings /manifesto.html. Accessed August 17, 2018.

Hansen, Anders. *Environment, Media, and Communication.* New York: Routledge, 2010.

Hansen, Anders, and Robert Cox, eds. *The Routledge Handbook of Environment and Communication.* New York: Routledge, 2015.

Hansen, Miriam. "The Mass Production of the Senses: Classical Cinema as Vernacular Modernism." *Modernism/Modernity* 6, no. 2 (1999): 59–77.

Haraway, Donna. "Anthropocene, Capitalocene, Chthulucene: Making Kin." *Environmental Humanities* 6 (2015): 159–65.

Harcourt, Bernard E. "Political Disobedience." In *Occupy: Three Inquiries in Disobedience.* Edited by W. J. T. Mitchell, Bernard E. Harcourt, and Michael Taussig. Chicago: University of Chicago Press, 2013, 45–91.

Hardt, Michael. "Globalization and Democracy." McMaster University Institute for Globalization and the Human Condition Working Paper Series, May 13, 2001.

———. "Two Faces of Apocalypse: Letter from Copenhagen." *Polygraph* 22 (2010): 265–74.

Hardt, Michael, and Antonio Negri. *Empire.* Cambridge, MA: Harvard University Press, 2000.

———. "Totality." In "'Subterranean Passages of Thought': Empire's Inserts." *Cultural Studies* 16, no. 2 (2002).

Harris, Bob. *The International Bank of Bob: Connecting Our World one $25 Kiva Loan at a Time.* New York: Walker and Company, 2013.

Harvey, David. *A Brief History of Neoliberalism.* New York: Oxford University Press, 2005.

———. *The New Imperialism.* New York: Oxford University Press, 2003.

Heidegger, Martin. *Discourse on Thinking.* Translated by J. M. Anderson and E. Hans Freund. New York: Harper, 1969.

———. "The Question Concerning Technology." In *The Question Concerning Technology and Other Essays.* Translated by William Lovitt. New York: Harper & Row, 1977, 3–35.

Heinberg, Richard. *The Party's Over: Oil, War, and the Fate of Industrial Societies.* New York: New Society, 2005.

Heintzman, Andrew, and Evan Solomon, eds. *Fueling the Future: How the Battle over Energy Is Changing Everything.* Toronto: Anansi, 2003.

Heise, Ursula. *Imagining Extinction: The Cultural Meanings of Endangered Species.* Chicago: University of Chicago Press, 2016.

Herder, Johann Gottfried. *Outlines of the History of Man.* Translated by T. Churchill. London: Johnson, 1800.

Herzog, Werner, dir. *Lessons of Darkness*. 1992; Troy, MI: Anchor Bay Entertainment, 2001. DVD.

Hirst, Paul, and Grahame Thompson. *Globalization in Question: The International Economy and the Possibilities of Governance*. Cambridge, UK: Wiley, 1996.

Hitchcock, Peter. *Imaginary States: Studies in Cultural Transnationalism*. Urbana: University of Illinois Press, 2003.

―――. "Oil in an American Imaginary." *New Formations* 69, no. 4 (2010): 81–97.

The Holy Bible (English Standard Version). Wheaton, IL: Crossway Bibles, 2001.

Homer-Dixon, Thomas, ed. *Carbon Shift: How Peak Oil and the Climate Crisis Will Change Canada (and Our Lives)*. Toronto: Random House, 2009.

Horkheimer, Max. *Critical Theory: Selected Essays*. New York: Continuum, 1975.

Horkheimer, Max, and Theodor Adorno. *Dialectic of Enlightenment*. Translated by John Cumming. New York: Continuum, 1988.

Howe, Cymene. "Anthropocenic Ecoauthority: The Winds of Oaxaca." *Anthropological Quarterly* 86, no. 1 (2012): 381–404.

Huber, Matthew. *Lifeblood: Oil, Freedom, and the Forces of Capital*. Minneapolis: University of Minnesota Press, 2013.

Humphreys, Macartan, Jeffrey D. Sachs, and Joseph E. Stiglitz. *Escaping the Resource Curse*. New York: Columbia University Press, 2007.

Illouz, Eva. *Cold Intimacies: The Making of Emotional Capitalism*. Cambridge, UK: Polity, 2007.

Innis, Harold. *The Fur Trade in Canada: An Introduction to Canadian Economic History*. Toronto: University of Toronto Press, 1977.

Jaccard, Mark. *Sustainable Fossil Fuels: The Unusual Suspect in the Quest for Clean and Enduring Energy*. Cambridge, UK: Cambridge University Press, 2006.

Jameson, Fredric. "A Brief Response." *Social Text* 19 (1987): 26–28.

―――. *The Cultural Turn: Selected Writings on the Postmodern, 1993–1998*. New York: Verso, 1998.

―――. *Fables of Aggression: Wyndham Lewis, the Modernist as Fascist*. Berkeley, CA: University of California Press, 1979.

―――. "Finance Capitalism and Culture." In *The Cultural Turn: Selected Writings on the Postmodern, 1983–1998*. New York: Verso, 1998, 136–61.

―――. "Globalization and Political Strategy." *New Left Review* 4 (2000): 49–68.

―――. "Metacommentary." *PMLA* 86, no. 1 (1971): 9–17.

―――. "Modernism and Imperialism." In *Nationalism, Colonialism, and Literature*. Co-authored by Terry Eagleton, Fredric Jameson, and Edward W. Said. Minneapolis: University of Minnesota Press, 1990, 43–66.

―――. "Notes on Globalization as a Philosophic Issue." In *The Cultures of Globalization*. Edited by Fredric Jameson and Masao Miyoshi. Durham, NC: Duke University Press, 1998.

―――. *The Political Unconscious: Narrative as a Socially Symbolic Act*. Ithaca, NY: Cornell University Press, 1981.

―――. "The Politics of Utopia." *New Left Review* 25 (January–February, 2004): 35–54.

―――. *Postmodernism, or, the Cultural Logic of Late Capitalism*. Durham, NC: Duke University Press, 1991.

―――. *The Seeds of Time*. New York: Columbia University Press, 1996.

———. "Third-World Literature in the Era of Multinational Capitalism." *Social Text* 15 (1986): 65–88.

Jamieson, Dale. *Reason in Dark Time: Why the Struggle against Climate Change Failed and What It Means for Our Future.* Oxford: Oxford University Press, 2014.

Jappe, Anselm. *Guy Debord.* Berkeley, CA: University of California Press, 1999.

Johnson, Bob. *Carbon Nation: Fossil Fuels in the Making of American Culture.* Lawrence: University Press of Kansas, 2014.

———. *Mineral Rites: An Archeology of the Fossil Economy.* Baltimore: Johns Hopkins University Press, 2018.

Jones, Christopher. "Petromyopia: Oil and the Energy Humanities." *Humanities* 5, no. 2 (2016), n.p.

———. *Routes of Power: Energy and Modern America.* Cambridge, MA: Harvard University Press, 2014.

Kapur, Geeta. "Globalisation and Culture." *Third Text* 11, no. 39 (1997): 21–38.

Karim, Lamia. "Demystifying Micro-credit: The Grameen Bank, NGOs, and Neoliberalism in Bangladesh." *Cultural Dynamics* 20, no. 1 (2008): 5–29.

———. *Microfinance and its Discontents: Women in Debt in Bangladesh.* Minneapolis: University of Minnesota Press, 2011.

Katz, Michael. *Why Don't American Cities Burn?* Philadelphia: University of Pennsylvania Press, 2012.

Kellner, Douglas. "Globalization and the Postmodern Turn." In *Globalization and Europe.* Edited by Roland Axtmann. London: Cassells, 1998, 23–42.

Klare, Michael. *Blood and Oil: The Dangers and Consequences of America's Growing Dependency on Imported Petroleum.* New York: Penguin, 2005

———. "Hooked! The Unyielding Grip of Fossil Fuels on Global Life." *TomDispatch. Com,* July 14, 2016. http://www.tomdispatch.com/post/176164/tomgram:_michael_klare,_fossil_fuels_forever/. Accessed August 17, 2018.

———. *Rising Powers, Shrinking Planet: The New Geopolitics of Energy.* New York: Holt, 2009.

Knechtel, John, ed. *FUEL.* Cambridge, MA: MIT Press, 2008.

Knight, Frank. *Risk, Uncertainty, and Profit.* Boston: Houghton Mifflin, 1921.

Knox, Hannah. "Footprints in the City: Models, Materiality, and the Cultural Politics of Climate Change." *Anthropological Quarterly* 87, no. 2 (2014): 405–29.

Kolbert, Elizabeth. "No Time." *New Yorker,* May 26, 2014. http://www.newyorker.com/magazine/2014/05/26/no-time. Accessed August 17, 2018.

Kovel, Joel. *The Enemy of Nature: The End of Capitalism or the End of the World?* New York: Zed Books, 2007.

Krugman, Paul. *The Return of Depression Economics and the Crisis of 2008.* New York: W. W. Norton, 2009.

Kumar, Amitava, ed. *World Bank Literature.* Minneapolis: University of Minnesota Press, 2003.

Kunstler, James Howard. *The Long Emergency: Surviving the Converging Catastrophes of the Twenty-First Century.* New York: Atlantic Monthly Press, 2005.

Lefsrud, Lianne, and Renate Meyer. "Science or Science Fiction? Experts' Discursive Construction of Climate Change." *Organizational Studies* 33, no. 11 (November 2012): 1477–1506.

LeMenager, Stephanie. *Living Oil: Petroleum and Culture in the American Century.* Oxford: Oxford University Press, 2013.

———. "Petro-Melancholia: The BP Blowout and the Arts of Grief." *Qui Parle: Critical Humanities and Social Sciences* 19, no. 2 (2011): 25–56.

Lemke, Thomas. "'The Birth of Bio-Politics': Michel Foucault's Lecture at the Collège de France on Neo-Liberal Governmentality." *Economy and Society* 30, no. 2 (2001): 190–207.

Lenin, V. I. *Imperialism: The Highest Stage of Capitalism.* London: Pluto Press, 1996.

Levant, Ezra. *Ethical Oil: The Case for Canada's Oil Sands.* Toronto: McClelland & Stewart, 2010.

Lloyd, David, and Paul Thomas. *Culture and the State.* New York: Routledge, 1998.

Logar, Ernst. *Invisible Oil.* Vienna: Springer-Verlag, 2010.

Lomborg, Bjørn. *The Skeptical Environmentalist: Measuring the Real State of the World.* Cambridge, UK: Cambridge University Press, 2001.

Luxemburg, Rosa. *The National Question.* Edited by Horace Davis. New York: Monthly Review Press, 1976.

Lyotard, Jean-Francois. *The Postmodern Condition: A Report on Knowledge.* Translated by Brian Massumi. Minneapolis: University of Minnesota Press, 1985.

Macdonald, Graeme. "The Resources of Fiction." *Reviews in Cultural Theory* 4, no. 2 (2013): 1–24.

Macdonald, Graeme, and Janet Stewart, eds. *Routledge Handbook of Energy Humanities.* London: Routledge, forthcoming 2019.

MacDuffie, Allen. *Victorian Literature, Energy, and the Ecological Imagination.* Cambridge, UK: Cambridge University Press, 2014.

Malm, Andreas. *Fossil Capital: The Rise of Steam Power and the Roots of Global Warming.* New York: Verso, 2016.

Marazzi, Christian. *The Violence of Financial Capitalism.* Los Angeles: Semiotext(e), 2010.

Marcuse, Herbert. *Negations: Essays in Critical Theory.* Translated by Jeremy J. Shapiro. London: Free Association Books, 1988.

Marriott, James, and Mika Minio-Paluello. *The Oil Road: Journeys from the Caspian Sea to the City of London.* New York: Verso, 2012.

Marsden, William. *Stupid to the Last Drop.* Toronto: Vintage Canada, 2008.

Marx, Karl. *Capital*, vol. 1. Translated by Ben Fowkes. New York: Penguin, 1976.

———. "Critique of the Gotha Program," In *The Marx-Engels Reader.* 2nd ed. Edited by Robert C. Tucker. New York: W. W. Norton, 1978, 525–41.

———. "The German Ideology: Part 1." In *The Marx-Engels Reader.* Edited by Robert C. Tucker. New York: W. W. Norton, 1978, 146–75.

———. *Grundrisse.* Harmondsworth: Penguin, 1973.

———. *Theories of Surplus Value*, 3 vols. Moscow: Progress Publishers, 1963–1971.

Mattelart, Armand. *Networking the World, 1794–2000.* Translated by Liz Carey-Libbrect and James A. Cohen. Minneapolis: University of Minnesota Press, 2000.

Mbembe, Achille. "At the Edge of the World: Boundaries, Territoriality, and Sovereignty in Africa." *Public Culture* 12, no. 1 (2000): 259–84.

McGonegal, Julie. "Post-Colonial Contradictions in Tsitsi Dangaremba's *Nervous Condition*: Toward a Reconsideration of Jameson's National Allegory." Unpublished manuscript.

McKillop, Andrew, and Sheila Newman, eds. *The Final Energy Crisis*. London: Pluto Press, 2005.

McLagan, Meg. "Introduction: Making Human Rights Claims Public." *American Anthropologist* 108, no. 1 (2006): 191–95.

McNeill, J. R. *Something New under the Sun: An Environmental History of the Twentieth-Century World*. New York: W. W. Norton, 2000.

McNeill, J. R., and Peter Engelke. *The Great Acceleration: An Environmental History of the Anthropocene since 1945*. Cambridge, MA: The Belknap Press, 2014.

McNeill, William. *Plagues and Peoples*. New York: Anchor, 1998.

Melville, Herman. *Moby Dick, or the Whale*. New York: Hendricks House, 1952.

Mies, Maria. *Patriarchy and Accumulation on a World Scale: Women in the International Division of Labor*. London: Zed Books, 2014.

Miller, Mark Crispin. *Boxed In: The Culture of TV*. Evanston, IL: Northwestern University Press, 1988.

Mitchell, Timothy. "Carbon Democracy." *Economy and Society* 38, no. 2 (2009): 399–432.

———. *Carbon Democracy: Political Power in the Age of Oil*. New York: Verso, 2011.

Miyoshi, Masao. "Ivory Tower in Escrow." *boundary* 227, no. 1 (2000): 7–50.

Monani, Salma. "Energizing Environmental Activism? Environmental Justice in Extreme Oil: The Wilderness and Oil on Ice." *Environmental Communication: A Journal of Nature and Culture* 2, no. 1 (2008): 119–27.

Montgomery, Scott L. *The Powers That Be: Global Energy for the Twenty-First Century and Beyond*. Chicago: University of Chicago Press, 2010.

Moore, Jason. *Capitalism in the Web of Life: Ecology and the Accumulation of Capital*. New York: Verso, 2015.

Moretti, Franco. "Conjectures on World Literature." *New Left Review* 1 (2000): 54–68.

———. *Modern Epic: The World-System from Goethe to García Márquez*. Translated by Quintin Hoare. New York: Verso, 1996.

Morris, Craig, and Arne Jungjohann. *Energy Democracy: Germany's Energiewende to Renewables*. Cham, Switzerland: Palgrave Macmillan, 2016.

Morton, Timothy. *Hyperobjects: Philosophy and Ecology after the End of the World*. Minneapolis: University of Minnesota Press, 2013.

Mouhout, Jean-François. "Past Connections and Present Similarities in Slave Ownership and Fossil Fuel Usage." *Climate Change* 105, no. 1–2 (March 2011): 329–55.

Munif, Adbul Rahman. *Cities of Salt* (Cities of Salt Trilogy, Vol. 1). New York: Vintage Books, 1987.

Murphy, D. J. "The Implications of the Declining Energy Return on Investment of Oil Production." *Phil. Trans. R. Soc. A.* 372, no. 2006 (2014): 1–19.

Nanos, Janelle. "The End of Ownership." *Boston Magazine*, May 2013. www.bostonmagazine.com/news/article/2013/04/30/end-ownership-sharing-conomy/. Accessed August 17, 2018.

National Business Incubation Association. "NBIA: Your Source for Knowledge and Networks in Business Incubation." *NBIA.org*, 2014. www.nbia.org/. Accessed August 17, 2018.

Natural Resources Canada. *Important Facts on Canada's Natural Resources*. Ottawa: Government of Canada, 2011.

Natural Resources Canada. *Pipelines Across Canada*. May 2016. https://www.nrcan .gc.ca/energy/infrastructure/18856. Accessed August 17, 2018.

Negarestani, Reza. *Cyclonopedia: Complicity with Anonymous Materials*. Praham, Australia: Re.Press, 2008.

Negri, Antonio. *Insurgencies: Constituent Power and the Modern State*. Translated by Maurizia Boscagli. Minneapolis: University of Minnesota Press, 2005.

Neuwirth, Robert. *Shadow Cities: A Billion Squatters in a New Urban World*. New York: Routledge, 2006.

———. *Stealth of Nations: The Global Rise of the Informal Economy*. New York: Pantheon, 2011.

Nikiforuk, Andrew. *The Energy of Slaves: Oil and the New Servitude*. Vancouver: Greystone Books, 2012.

———. *Saboteurs: Wiebo Ludwig's War against Big Oil*. Toronto: Macfarlane Walter and Ross, 2001.

O'Brien, Susie, and Imre Szeman. *Content Providers of the World Unite! The Cultural Politics of Globalization*. Working Paper GHC 03/3. Hamilton, ON: Institute on Globalization and the Human Condition Working Paper Series, McMaster University, 2003.

Ogden, Joan. "High Hopes for Hydrogen." *Scientific American* 295, no. 3 (2006): 94–101.

O'Malley, Martin. *Past and Future Land: An Account of the Berger Inquiry into the Proposed Mackenzie Valley Pipeline*. Toronto: Peter Martin Associates, 1976.

Oosthoek, Jan, and Barry K. Gills. "Humanity at the Crossroads: The Globalization of Environmental Crisis." *Globalizations* 2, no. 3 (2005): 283–91.

Orestes, Naomi. "The Scientific Consensus on Climate Change." *Science* 306, no. 5702 (Dec. 3, 2004): 1686.

Peck, Jamie. "The Creativity Fix." *Eurozine*, June 28, 2007. https://www.eurozine.com /the-creativity-fix/. Accessed August 17, 2018.

Poovey, Mary. "The Twenty-First Century University: What Price Economic Viability?" *differences: A Journal of Feminist Cultural Studies* 12, no. 1 (2001): 1–16.

Pope Francis. *Encyclical on Climate Change and Inequality: On Care for our Common Home*. Brooklyn: Melville House, 2015.

Porter, Michael E. "The Competitive Advantage of the Inner City." *Harvard Business Review* 73, no. 3 (May-June 1995): 55–71.

Prahad, C. K. *The Fortune at the Bottom of the Pyramid: Eradicating Poverty through Profits*. Philadelphia: Wharton School, 2006.

Pratt, Mary Louise. "Planetary Longings: Sitting in the Light of the Great Solar TV." In *World Writing: Poetics, Ethics, Globalization*. Edited by Mary Gallagher. Toronto: University of Toronto Press, 2008, 207–23.

Princen, Thomas, Jack P. Manno, and Pamela L. Martin, eds. *Ending the Fossil Fuel Era*. Cambridge, MA: MIT Press, 2015.

Radhakrishnan, R. "Postmodernism and the Rest of the World." In *The Pre-Occupation of Postcolonial Studies*. Edited by Fawzia Afzal-Khan and Kalpana Seshadri-Crooks. Durham, NC: Duke University Press, 2000, 33–70.

Rancière, Jacques. "Comments and Responses." *Theory and Event* 6, no. 4 (2003), n.p.

———. *The Emancipated Spectator*. London: Verso, 2010.

Raunig, Gerald. *Art and Revolution: Transversal Activism in the Long Twentieth Century*. Translated by Aileen Derieg. Los Angeles: Semiotext(e), 2007.

Rawls, John. "Justice as Fairness: Political not Metaphysical." *Philosophy and Public Affairs* 14, no. 3 (1985): 223–51.

Readings, Bill. *The University in Ruins*. Cambridge, MA: Harvard University Press, 1996.

Renshaw, Edward. "The Substitution of Inanimate Energy for Animal Power." *Journal of Political Economy* 71, no. 3 (1963): 284–92.

Retort. *Afflicted Powers: Capital and Spectacle in a New Age of War*. New York: Verso, 2005.

———. "An Exchange on *Afflicted Powers: Capital and Spectacle in a New Age of War*." *October* 115 (2006): 3–12.

Roberts, Paul. *The End of Oil: On the Edge of a Perilous New World*. New York: Mariner Books, 2005.

Ross, Andrew. "The Mental Labour Problem." *Social Text* 63 (2000): 1–31.

———. *Nice Work If You Can Get It: Life and Labor in Precarious Times*. New York: New York University Press, 2009.

Rubin, Jeff. *The End of Growth*. Toronto: Random House, 2012.

Sandberg, Sheryl. *Lean In: Women, Work, and the Will to Lead*. New York: Knopf, 2013.

Sassen, Saskia. "Spatialities and Temporalities of the Global: Elements of a Theorization." *Public Culture* 12, no. 1 (2000): 215–32.

Scarry, Elaine. *On Beauty and Being Just*. Princeton, NJ: Princeton University Press, 1999.

Schor, Juliet. *The Overworked American: The Unexpected Decline of Leisure*. New York: Basic Books, 1993.

———. "Towards a New Politics of Consumption." In *The Consumer Society Reader*. Edited by Juliet B. Schor and Douglas Holt. New York: The New Press, 2000, 446–62.

Schumpeter, Joseph. *The Theory of Economic Development: An Inquiry into Profits, Capital, Credit, Interest, and the Business Cycle*. Cambridge, MA: Harvard University Press, 1934.

Seshadri-Crooks, Kalpana. "At the Margins of Postcolonial Studies: Part I." In *The Pre-Occupation of Postcolonial Studies*. Edited by Fawzia Afzal-Khan and Kalpana Seshadri-Crooks. Durham, NC: Duke University Press, 2000, 3–23.

Simmons, Matthew. *Twilight in the Desert: The Coming Saudi Oil Shock and the World Economy*. Hoboken, NJ: Wiley, 2005.

Sinclair, Upton. *Oil!* New York: Grosset & Dunlap, 1927.

Sisson, Natalie. *The Suitcase Entrepreneur: Create Business and Adventure in Life*. Tonawhai Press, 2013.

Sloterdijk, Peter. *Critique of Cynical Reason*. Translated by Michael Eldred. Minneapolis: University of Minnesota Press, 1987.

Smil, Vaclav. *Energy Transitions*. New York: Praeger, 2010.

———. "World History and Energy." In *Encyclopedia of Energy* Vol. 2. Edited by J. Vutler Cleveland. 6 vols. Amsterdam: Elsevier, 2004, 549–61.

Smith, Jessica, and Mette High. "Exploring the anthropology of energy: Ethnography, energy, and ethics." *Energy Research and Social Science* 30 (2017): 1–6.

Smith, Neil. *The Endgame of Globalization*. New York: Routledge, 2004.

Socolow, Robert H., and Stephen W. Pacala. "A Plan to Keep Carbon in Check." *Scientific American* 295, no. 3 (2006): 50–57.

Southwood, Ivor. *Non-Stop Inertia*. Winchester, UK: Zero Books, 2010.

Spivak, Gayatri Chakravorty. "Can the Subaltern Speak? Speculations on Widow Sacrifice." *Wedge* 7/8 (1985): 120–30.

———. *Critique of Postcolonial Reason*. Cambridge, MA: Harvard University Press, 1999.

———. *Death of a Discipline*. New York: Columbia University Press, 2003.

Sprinker, Michael. "The National Question: Said, Ahmad, Jameson." *Public Culture* 6 (1993): 3–29.

Steiner, Wendy. *Venus in Exile: The Rejection of Beauty in Twentieth-Century Art*. New York: Free Press, 2001.

Stevens, Jacob. "Monetized Ecology." *New Left Review* 16 (2002): 143–51.

Stix, Gary. "A Climate Repair Manual." *Scientific American* 295, no. 3 (2006): 46–49.

Stoekl, Allan. *Bataille's Peak: Energy, Religion, and Postsustainability*. Minneapolis: University of Minnesota Press, 2007.

———. "Unconventional Oil and the Gift of the Undulating Peak." *Imaginations: Journal of Cross-Cultural Image Studies* 3, no. 2 (2012): 35–45.

Streeck, Wolfgang. "How Will Capitalism End?" *New Left Review*, 87 (2014): 35–64.

Szeman, Imre. "Belated or Isochronic?: Canadian Writing, Time and Globalization." *Essays on Canadian Writing* 71 (2000): 145–53.

———. "Crude Aesthetics: The Politics of Oil Documentaries." *Journal of American Studies* 46 no. 2 (2012): 423–39.

———. "The Cultural Politics of Oil: On *Lessons of Darkness* and *Black Sea Files*." *Polygraph* 22 (2010): 3–15.

———. "On Energopolitics." *Anthropological Quarterly* 87, no. 2 (2014): 453–64.

———. "On the Politics of Extraction." *Cultural Studies* 31, no. 2 (2017): 440–47.

———. "System Failure: Oil, Futurity, and the Anticipation of Disaster." *The South Atlantic Quarterly* 106 no. 4 (2007): 805–23.

———. *Zones of Instability: Literature, Postcolonialism, and the Nation*. Baltimore: Johns Hopkins University Press, 2003.

Szeman, Imre, and Dominic Boyer, eds. *Energy Humanities: An Anthology*. Baltimore: Johns Hopkins University Press, 2017.

Szeman, Imre, Jennifer Wenzel, and Patricia Yaeger, eds. *Fueling Culture: 101 Words for Energy and Environment*. New York: Fordham University Press, 2017.

Taine, Hippolyte. *History of English Literature*. Translated by H. Van Laun. New York: F. Ungar, 1965.

Tardy, Jaime. *The Eventual Millionaire: How Anyone Can Be an Entrepreneur*. Hoboken, NJ: Wiley, 2014.

Taylor, Peter. *Modernities: A Geohistorical Interpretation*. Minneapolis: University of Minnesota Press, 1999.

Terranova, Tiziana. "Free Labor: Producing Culture for the Digital Economy." *Social Text* 18, no. 2 (2000): 33–58.

Therborn, Goran. "Into the 21st Century: The New Parameters of Global Politics." *New Left Review* 10 (2001): 87–110.

Tokumitsu, Miya. "In the Name of Love." *Jacobin* 13 (2014). www.jacobinmag.com/2014/01/in-the-name-of-love/. Accessed August 17, 2018.

Tomlinson, John. *Cultural Imperialism*. Baltimore: Johns Hopkins University Press, 1991.

———. *Globalization and Culture*. Chicago: University of Chicago Press, 1999.

Trainer, Ted. "The Simpler Way." In *The Final Energy Crisis*. Edited by Andrew McKillop and Sheila Newman. London: Pluto Press, 2005, 279–88.

Venkatesh, Sudhir. *Off the Books: The Underground Economy of the Urban Poor*. Cambridge, MA: Harvard University Press, 2008.

Virno, Paolo. "The Dismeasure of Art: An Interview with Paolo Virno." *Open* 17 (2009). https://chtodelat.org/wp-content/uploads/2009/10/Virno_Dismeasure.pdf. Accessed August 17, 2018.

———. *A Grammar of the Multitude: For an Analysis of Contemporary Forms of Life*. New York: Semiotext(e), 2004.

Vishmidt, Marina. "Mimesis of the Hardened and Alienated: Social Practice as Business Model." In *Disrupting Business*. Edited by Tatiana Bazzichelli and Geoff Cox. New York: Autonomedia, 2013, 91–109.

———. "Permanent Reproductive Crisis: An Interview with Silvia Federici." *Mute*, March 7, 2013. www.metamute.org/editorial/articles/permanent-reproductive-crisis-interview-silvia-federici. Accessed August 17, 2018.

Vital, Anna. "Everyone Will Have to Become an Entrepreneur." *Funders and Founders*, Feb. 20, 2013. www.fundersandfounders.com/everyone-will-have-to-become-an-entrepreneur/. Accessed August 17, 2018.

Walsh, Shannon, dir. *H₂Oil*. 2009. Loaded Pictures, 2009. DVD.

Watson, James L., ed. *Golden Arches East: McDonald's in East Asia*. Stanford, CA: Stanford University Press, 1997.

Wellum, Caleb. "The Ambivalent Aesthetics of Oil: Project Documerica and the Energy Crisis in 1970s America." *Environmental History* 22, no. 4 (Oct. 2017): 723–32.

Wenzel, Jennifer. "Consumption for the Common Good? Commodity Biography Film in an Age of Postconsumerism." *Public Culture* 23, no. 3 (Fall 2011): 572–602.

———. "Introduction." In *Fueling Culture: 101 Words for Energy and Environment*. Edited by Imre Szeman, Jennifer Wenzel, and Patricia Yaeger. New York: Fordham University Press, 2017, 1–16.

———. "Petro-Magic-Realism: Towards a Political Ecology of Nigerian Literature." *Postcolonial Studies* 9, no. 4 (2006): 449–64.

Whyte, William. *The Organization Man*. New York: Simon and Schuster, 1956.

Williams, Colin C., and Sara Nadin. "Entrepreneurship and the Informal Economy: An Overview." *Journal of Developmental Entrepreneurship* 15, no. 4 (2010): 361–78.

Williams, Raymond. "Culture Is Ordinary." In *Cultural Theory: An Anthology*. Edited by Imre Szeman and Timothy Kaposy. Oxford: Wiley-Blackwell, 2010, 53–59.

———. *Marxism and Literature*. Oxford: Oxford University Press, 1977.

———. *Problems in Materialism and Culture: Selected Essays*. New York: Verso, 1980.

Wilson, Sheena, Adam Carlson, and Imre Szeman, eds. *Petrocultures: Oil, Politics, Culture*. Montreal-Kingston: McGill-Queen's University Press, 2017.

Wood, John H., Gary R. Long, and David F. Morehouse. "Long-Term World Oil Supply Scenarios: The Future Is Neither as Bleak or Rosy as Some Assert." *Energy Information Administration*. Aug. 18, 2004. http://large.stanford.edu/publications /coal/references/wood/. Accessed January 16, 2019.

———. "2000 U.S. Geological Survey World Petroleum Assessment." *United States Geological Survey*. http://pubs.usgs.gov/dds/dds-060/. Accessed January 16, 2019.

Yaeger, Patricia. "Editor's Column: Literature in the Ages of Wood, Tallow, Coal, Whale Oil, Gasoline, Atomic Power, and Other Energy Sources." *PMLA* 126, no. 2 (2011): 305–26.

Yergin, Daniel. "Ensuring Energy Security." *Foreign Affairs* 85, no. 2 (2006): 67–82.

———. *The Prize: The Epic Quest for Oil, Money, and Power*. New York: Free Press, 2008.

———. *The Quest: Energy, Security, and the Remaking of the Modern World*. New York: Penguin, 2011.

Yúdice, George. *The Expediency of Culture*. Durham, NC: Duke University Press, 2003.

Yunus, Muhammed. *Banker to the Poor: Micro-Lending and the Battle against World Poverty*. New York: Public Affairs, 1999.

Žižek, Slavoj. *In Defense of Lost Causes*. New York: Verso, 2008.

———. "Introduction: The Spectre of Ideology." In *Mapping Ideology*. Edited by Slavoj Žižek. New York: Verso, 1994, 1–33.

———. *On Belief*. London: Routledge, 2003.

———. *The Sublime Object of Ideology*. New York: Verso, 2008.

———. *Welcome to the Desert of the Real*. New York: Verso, 2002.

Sources and Permissions

Several of the essays included in this volume were previously published and are reproduced with permission. Changes have been made to the original in each case.

Chapter 1: "Who's Afraid of National Allegory? Jameson, Literary Criticism, Globalization." *The South Atlantic Quarterly* 100, no. 3 (2001): 801–25. Copyright © 2001 Duke University Press. All rights reserved. Reprinted by permission of the publisher (www.dukeupress.edu).

Chapter 2: "Culture and Globalization, or, The Humanities in Ruins." *CR: The New Centennial Review* 3, no. 2 (2003): 91–115. Copyright © 2003 Michigan State University. Reprinted by permission of the publisher.

Chapter 3: "Globalization, Postmodernism, and (Autonomous) Criticism." *Cultural Autonomy: Frictions and Connections*. Edited by Petra Rethmann, Imre Szeman, and William Coleman. Vancouver: University of British Columbia Press, 2010, 66–85. Copyright © 2010 University of British Columbia Press. All rights reserved. Reprinted by permission of the publisher.

Chapter 4: "System Failure: Oil, Futurity, and the Anticipation of Disaster." *The South Atlantic Quarterly* 106, no. 4 (2007): 805–23. Copyright © 2007 Duke University Press. All rights reserved. Reprinted by permission of the publisher (www.dukeupress.edu).

Chapter 5: "Neoliberals Dressed in Black; or, the Traffic in Creativity," *English Studies in Canada* 36, no. 1 (2010): 15–36. Copyright © 2010 Association of Canadian College and University Teachers of English. Reprinted by permission of the publisher.

Chapter 6: "The Cultural Politics of Oil: On *Lessons of Darkness* and *Black Sea Files*." *Polygraph* 22 (2010): 3–15. Copyright © 2010 Polygraph. Reprinted by permission of the publisher.

Chapter 7: "Crude Aesthetics: The Politics of Oil Documentaries." *Journal of American Studies* 46, no. 2 (2012): 423–39. Copyright © 2012 Cambridge University Press. Reprinted by permission of the publisher.

Chapter 8: "How to Know about Oil: Energy Epistemologies and Political Futures." *Journal of Canadian Studies* 47, no. 3 (2013): 145–68. Copyright © 2013 Journal of Canadian Studies. Reprinted by permission of University of Toronto Press (www.utpjournals.com).

Chapter 9: "Entrepreneurship as the New Common Sense." *The South Atlantic Quarterly* 114, no. 3 (2015): 471–90. Copyright © 2015 Duke University Press. All rights reserved. Reprinted by permission of the publisher (www.dukeupress.edu).

Chapter 10: "Conjectures on World Energy Literature." *Journal of Postcolonial Writing* 53, no. 2 (2017): 1–12. https://doi.org/10.1080/17449855.2017.1337672. Copyright © 2017 Taylor and Francis. Reprinted by permission of the publisher.

Chapter 12: "On the Politics of Region." *e-flux Architecture* (2018), Online. Reprinted by permission of the publisher.

Index

Index

terrorism, 52, 108
Texaco, 94, 160, 161, 165
Thatcher, Margaret, 80
Thiel, Peter, 207, 218n26
Thomas, Paul, 57
Tide Water, 247–48
Tomlinson, John, 55
Toronto, Ontario, 1, 114, 217n8
totality, 25, 43, 44, 46n38
trade, 5, 39, 49, 56, 77, 83, 84, 96,
    98, 181, 190, 220, 224, 249, 263,
    265
Trans Mountain Pipeline, 239, 240,
    242–43, 253
Trans-Arabian Pipeline, 249
TransCanada Corporation,
    242–44, 261
Trump administration, 6–7, 13,
    239, 251, 261

U

UNESCO, 134n4
United Nations, 236n15, 255
  Climate Change Conference
    (COP), 170, 244
  Intergovernmental Panel on
    Climate Change, 3
United States, 26, 40–41, 46n38,
    47n53, 51, 116, 119, 123, 129,
    184, 204, 208, 217n17
  Advanced Energy Initiative,
    97–101
  Bush administration, 97–98, 157,
    191
  Council on Foreign Relations,
    111n12
  and energy, 6, 95, 98, 137, 163,
    178, 189, 197n16, 228–29,

    236n12, 238–39, 242, 245,
    247, 250
  Gulf War, 142, 147
  invasion of Afghanistan, 192
  invasion of Iraq, 98, 100, 107,
    108, 111n12, 142, 192, 193
  and the Middle East, 6–7, 97,
    111n12, 228, 229–30
  Obama administration, 239, 251,
    260
  Trump administration, 6–7, 13,
    239, 251, 261
universities. *See* academia
University of Alberta, 216n7
University of Toronto, 114, 217n7
University of Wisconsin–
    Milwaukee, 216–217n7

V

Veiga, Ricardo Reis, 160
Virno, Paolo, 132, 133, 136n48,
    194
Vishmidt, Marina, 215, 217n8,
    218–19n33
Vital, Anna, 203

W

Wallerstein, Immanuel, 93
Walsh, Shannon, 154, 156–57, 159,
    162–64, 165–67, 167, 168
Watts, Peter, 87
Wente, Margaret, 69n4
Wenzel, Jennifer, 8–9, 227
Wertheimer, Max, 135n15
Whyte, William, 202
Wilder, Billy, 45n17

297